Tourism, Performance and the Everyday

Tourism has become increasingly 'exotic', a process made possible by low-cost charter tourism and cheaper air tickets. Faraway and ever more 'exotic' holidays are becoming widespread and within reach as destinations make their entry into the mass tourism market. Strolls through the bazaars of Istanbul and cruises on the Nile are packaged into the sea, sand and sun culture of traditional forms of organized mass tourism. At the same time new technologies weave the fabric of tourism and everyday life even closer, circulating images, information and objects between them. Taking off from this observation, *Tourism, Performance and the Everyday* invites readers to follow the flows of tourist desires, objects, meanings, photographs, fears, dreams and memories weaving together the spaces of and between Western Europe, Turkey and Egypt.

This book carefully analyses the cultural and social impacts of mass-tourist experiences of 'exotic' places on the wider aspects of everyday life. It treats mass tourism as a cultural phenomenon that feeds into the practices and networks of people's everyday lives rather than as an isolated, trivial or 'exotic' event. Chapters cover topics such as the material cultures of tourism, mobilizing the Orient and the afterlife of tourism. The argument traces how these impacts are mediated by various mobilities between home and away through innovative mobile and ethnographic research methods at tourist destinations and the homes of tourists. The book contains analysis of diaries, photographs, blogs and web photo-sharing sites, participant observation of performing tourists and 'home ethnographies' of the afterlife of tourist photographs, souvenirs and memories.

In doing this, the book traces out the multiple interconnections and mobilities between everyday spaces and leisure spaces as well as the multiple ways in which the Orient is consumed on holiday and at home. The book will appeal to a wide readership among students, researchers and educators within the social and cultural sciences studying, researching and teaching theories and methods of tourism, Orientalism and cultural encounters as well as broader issues of leisure, consumption, mobilities studies and everyday life.

Michael Haldrup is a lecturer in Geography at Roskilde University, Denmark, and was a Visiting Research Fellow at Lancaster University, UK, 2007. His main research interest is in tourism, place and everyday life, with a longstanding interest in social theory and spatial relations. He has written extensively on issues relating to mobility, place, identity, cultural industries, heritage and tourism.

Jonas Larsen is a lecturer in Geography at Roskilde University, Denmark. He received his PhD degree in Cultural Geography in 2004. He is interested in mobility, tourism and media and has published 16 refereed articles on tourism, geography, mobility and media in international journals and co-authored two books.

Contemporary Geographies of Leisure, Tourism and Mobility

Series Editor: C. Michael Hall

Professor at the Department of Management, College of Business & Economics, University of Canterbury, Private Bag 4800, Christchurch, New Zealand

The aim of this series is to explore and communicate the intersections and relationships between leisure, tourism and human mobility within the social sciences.

It will incorporate both traditional and new perspectives on leisure and tourism from contemporary geography, e.g. notions of identity, representation and culture, while also providing for perspectives from cognate areas such as anthropology, cultural studies, gastronomy and food studies, marketing, policy studies and political economy, regional and urban planning, and sociology, within the development of an integrated field of leisure and tourism studies.

Also, increasingly, tourism and leisure are regarded as steps in a continuum of human mobility. Inclusion of mobility in the series offers the prospect to examine the relationship between tourism and migration, the sojourner, educational travel, and second home and retirement travel phenomena.

The series comprises two strands:

Contemporary Geographies of Leisure, Tourism and Mobility aims to address the needs of students and academics, and the titles will be published in hardback and paperback. Titles include:

The Moralisation of Tourism
Sun, sand . . . and saving the world?
Jim Butcher

The Ethics of Tourism Development
Mick Smith and Rosaleen Duffy

Tourism in the Caribbean
Trends, development, prospects
Edited by David Timothy Duval

Qualitative Research in Tourism
Ontologies, epistemologies and methodologies
Edited by Jenny Phillimore and Lisa Goodson

The Media and the Tourist Imagination
Converging cultures
Edited by David Crouch, Rhona Jackson and Felix Thompson

Tourism and Global Environmental Change
Ecological, social, economic and political interrelationships
Edited by Stefan Gössling and C. Michael Hall

Cultural Heritage of Tourism in the Developing World
Dallen J. Timothy and Gyan Nyaupane

Understanding and Managing Tourism Impacts
Michael Hall and Alan Lew

Routledge Studies in Contemporary Geographies of Leisure, Tourism and Mobility is a forum for innovative new research intended for research students and academics, and the titles will be available in hardback only. Titles include:

1. **Living with Tourism**
   Negotiating identities in a Turkish village
   Hazel Tucker

2. **Tourism, Diasporas and Space**
   Edited by Tim Coles and Dallen J. Timothy

3. **Tourism and Postcolonialism**
   Contested discourses, identities and representations
   Edited by C. Michael Hall and Hazel Tucker

4. **Tourism, Religion and Spiritual Journeys**
   Edited by Dallen J. Timothy and Daniel H. Olsen

5. **China's Outbound Tourism**
   Wolfgang Georg Arlt

6. **Tourism, Power and Space**
   Edited by Andrew Church and Tim Coles

7. **Tourism, Ethnic Diversity and the City**
   Edited by Jan Rath

8. **Ecotourism, NGOs and Development**
   A critical analysis
   Jim Butcher

9. **Tourism and the Consumption of Wildlife**
   Hunting, shooting and sport fishing
   Edited by Brent Lovelock

10. **Tourism, Creativity and Development**
    Edited by Greg Richards and Julie Wilson

Tourism, Performance and the Everyday

Consuming the Orient

Michael Haldrup and
Jonas Larsen

Routledge
Taylor & Francis Group

LONDON AND NEW YORK

First published 2010
by Routledge
2 Park Square, Milton Park, Abingdon, Oxon, OX14 4RN

Simultaneously published in the USA and Canada
by Routledge
270 Madison Avenue, New York, NY 10016

*Routledge is an imprint of the Taylor & Francis Group, an informa
business*

© 2010 Michael Haldrup and Jonas Larsen

Typeset in Times New Roman by Prepress Projects Ltd, Perth, UK
Printed and bound by the MPG Books Group in the UK

British Library Cataloguing in Publication Data
A catalogue record for this book is available from the British Library

Library of Congress Cataloguing in Publication Data
Haldrup, Michael.
Tourism, performance, and the everyday : consuming the Orient / Michael
Haldrup and Jonas Larsen.
p. cm.
Includes bibliographical references and index.
1. Tourism – Orient. 2. Tourism – Social aspects – Orient. I. Larsen, Jonas.
II. Title.
G155.O75H35 2009
306.4'819095–dc22
2009003564

ISBN 13: 978–0–415–46713–1 (hbk)
ISBN 13: 978–0–203–87393–9 (ebk)

ISBN 10: 0–415–46713–6 (hbk)
ISBN 10: 0–203–87393–9 (ebk)

Contents

Figures

Acknowledgements

The writing of this book was made possible by a grant from the Danish Research Council for Society and Business and a contribution from the Department of Environmental, Social and Spatial Change (ENSPAC), Roskilde University (2005–2008). In the preparation of the final manuscript we benefited from the help of Andrew Crabtree and Ditte Junge in proofreading, Ritta Juel Bitsch in preparing the maps, and Matthew Lequick, who provided graphic assistance.

This book is the product of a collaborative work process. Chapters 1, 3, 4, 6 and 8 were jointly written; Chapters 2 and 7 were written by Jonas Larsen and Chapters 5 and 9 by Michael Haldrup. Some of the work presented has formed part of various other writings by us. Most notably, earlier and shorter versions of Chapter 2 and Chapter 4 have been published in *Leisure Studies* (Haldrup and Larsen, 2006; Larsen, 2006). Chapter 7 draws on an article published in *Mobilities* (Larsen, 2008) and Chapter 9 on a chapter in *Cultures of Mass Tourism* (Haldrup, 2009).

During the process of writing we have presented the arguments and material in this book at conferences and seminars in Scandinavia and Great Britain, and we would like to thank Joan Amer, Keld Buciek, Javier Caletrio, Kingsley Dennis, Bülent Diken, Antje Gimmler, Derek Gregory, Michael Hviid Jacobsen, Lasse Koefoed, Jennie Germann Molz and Kirsten Simonsen for inspiring cooperation, comments and contributions to our work. We have greatly benefited from the encouragement and (as always) constructive suggestions from Jørgen Ole Bærenholdt, Roskilde University, and John Urry, Lancaster University.

Jonas is grateful that Johanna participated on a fieldwork trip to Alanya; Elliott is still happy that he can play with his father at the pool when he is at work. Michael wants to thank his family, Bjarke and Wibeke, for (once more) acting as his fieldwork cover, this time in Egypt. He is also grateful to them for spending half a year with him in the north-west of England and to John Urry for hosting him at the Sociology Department, Lancaster University, from January to June 2007.

Finally, both authors want to thank the people who have contributed, and allowed us to use, material for this book through interviews and diaries, and by opening up their homes for us.

1 Performing tourism, performing the Orient

Introduction

One of the most striking features of contemporary social life is the persistent presence of the Other in consumption and everyday life. The most remote corners of the globe now seem within daily reach when we work, shop, dine, watch television, surf and participate in the internet or walk the streets. Even the most spatially bounded forms of local social life and identity in the neighbourhood community or within the walls of territorial nation states are in part enacted on the premise of global flows, networks and scapes, and legitimized with reference to the perceived threats of the mobility of migrants, crime, terror, diseases, pollution, goods, capital and so on. In a sense we have all become 'everyday cosmopolitans' (Beck, 2006). Whether or not people turn their back on it, the Other emerges in the midst of the everyday through a variety of such mobilities. In contemporary discourse 'the Orient' has very much emerged as the epitome of Otherness to West European and North American culture, economy and society. To a degree not experienced before, the West and the Orient are woven together in webs of mobilities, corporeal, imaginative and virtual. The Orient has travelled to the West through the immigration of humans and the travelling of objects, signs and ideas. In many European towns and most cities, the Orient is within reach and consumable: one can see, touch, hear, eat, smell and click on it.

Although the Orient is part of Western home geographies, Westerners increasingly jump on jet planes to consume 'the real thing'. As a result of low-cost charter tourism and cheaper air tickets, many consume the places, people and cultures of the Orient at first hand, and perform connectedness with or detachment from these when performing tourism. Despite recurrent terror bombings at tourist destinations, global tourism continues its massive growth, and the Middle East countries are among the prime drivers (World Tourist Organization, 2008). Strolls through the bazaars of Istanbul and cruises on the Nile are packaged into the sea, sand and sun culture of traditional forms of organized mass tourism. Safe but beaten tracks are replaced with exotic geographies of Oriental Otherness.

This present book is concerned with rethinking the nature of tourism consumption. Globalization theorists address the changing spatio-temporal relationships of the contemporary world under labels such as 'time–space compression'

(Harvey, 1989), cultural hybridity (Bhabha, 1994) and global flows (Castells, 1996, 2000). Instead of thinking of 'the global' as something that merely adds to, extends, enriches or threatens our everyday lives and capabilities to consume, we suggest refocusing on the interdependent mobilities that *transform* everyday performances from within. As the global moves within reach, it also emerges as a condition for the mundane practices, minutiae and performances of day-to-day routines and interactions. We become intimate with the global, and the intimate fabrics of our daily lives become global. Touristic consumption is a particularly interesting example of this globalization 'from within'. As tourist travel becomes widespread, ever more people and places are drawn into the cultural-economic circuit of the tourist industry, and as the global cultural industry fills our everyday spaces with touristic objects and signs so the world can be consumed from afar. Conventionally, tourism theories define tourism in contrast to the everyday, but this book examines how tourism and the everyday complexly converge on many fronts. It shows how there is a touristification of everyday life and especially how everydayness informs tourists' performances. Tourist travel is not only about consuming Otherness but also about (re)producing social relations within family and friendship networks. Souvenirs, videos and tourist photographs decorate and become part of our bodies, living rooms and photograph albums while blogs, home pages, social networking sites such as Facebook and multimedia-sharing sites such as Flickr (photos) and YouTube (video), made possible by the 'Web 2.0 revolution', have made the internet more open, collaborative and participatory.

A central source of inspiration for this book is the increased awareness of practices, materials and ethnographic-inspired methodologies within parts of consumption studies in which attention has been redirected from semiotic studies of goods' symbol values ('sign-value') towards material readings of everyday consumption practices and the affordances, or materiality, of objects ('use-value') (Warde, 2005; Watson, 2008). This has produced rich ethnographies of consumption practices in shopping malls (Miller *et al.*, 1998), second-hand cultures (Gregson and Crewe, 2003) and private homes (Miller, 2001a), thus uncovering the many different everyday consumption practices and the variety of spaces in which they take place. By doing this, consumption studies has located itself within the material culture of people's everyday lives and explored the roles that mundane (and material) consumer objects play in our everyday lives, as well as the technological, social, cultural and political trajectories that enable particular practices and objects to be enacted. Although practice is concerned with doings, practices cannot be reduced to what people do. Equally, there is no such thing as 'just' doing. Instead, doings are doings *and* performances, 'shaped by and constitutive of the complex relations – of materials, knowledges, norms, meanings and so on – which comprise the practice-as-entity' (Shove *et al.*, 2007: 13), and in this book we approach tourist consumption though the lens of performance. In this way we want to locate it within the recent 'performance turn' in tourism studies, hence the title of the book: *Tourism, Performance and the Everyday: Consuming the Orient.*

The performance turn in tourist studies

In some ways this book is an extension of our book *Performing Tourist Places* (Bærenholdt *et al.,* 2004), in which we studied the production and consumption of tourist places in Denmark through the lens of performance. That book is framed by and itself constitutive of a performance turn that can be traced back to tourism theory and research in the 1990s. This turn is evident in research books and articles (Edensor, 1998; Desmond, 1999; Löfgren, 1999; Perkins and Thorns, 2001; Coleman and Crang, 2002; Bærenholdt *et al.,* 2004, 2008; Lew *et al.*, 2004; Malam, 2004; Sheller and Urry, 2004; Doorne and Atejlevic, 2005; Ritchie *et al.,* 2005; Weave, 2005; Hannam, 2006; Minca and Oakes, 2006; Quinn, 2007; Ek *et al.*, 2008; Obrador Pons *et al.*, 2010). Given the danger of neglecting differences, it is justifiable to speak of a turn because the above publications share important departures from classical mainstream tourism theories. Much cultural tourism research has been concerned with how tourists are drawn to and experience – sense and represent – destinations, and the performance turn continues in that direction. But it also redirects tourism theory in some important ways. Although we believe that tourist studies have gained much from this turn, we also think that there is scope for theoretical and methodological progress and refinement. The performance turn is formed in opposition to the 'tourist gaze' and other representational approaches privileging the eye (e.g. MacCannell, 1976/1999; Shields, 1991; Dann, 1996a; Selwyn, 1996) by arguing that 'tourism demands new metaphors based more on being, doing, touching and seeing rather than just "seeing" ' (Perkins and Thorns, 2001: 189; see also Cloke and Perkins, 1998). In particular, John Urry's notion of the tourist gaze (Urry, 1990, 2002a) has been influential in portraying the tourist experience as a visual experience. The 'tourist gaze' suggests that people travel to destinations because of their striking visual qualities. In contrast, the performance turn highlights how tourists experience places in multisensuous ways that can involve multiple bodily sensations: touching, smelling, hearing, and so on. Tourists encounter cities and landscapes through corporeal proximity as well as distanced contemplation. Metaphorically speaking, in addition to looking at stages, tourists step into them and enact them corporeally.

Cultural accounts of tourism have often been trapped within a representational world of place myths and semiological readings of space (Hughes, 1998). The performance turn destabilizes semiotic readings in which places and objects are seen as signifying social constructs that can be unveiled through authoritative cultural readings rather than in terms of how they are used and lived with in practice. By shifting the focus to ontologies of acting and doing (Franklin and Crang, 2001) the corporeality of tourist bodies and their creative potentials, as well as the significance of technologies and the material affordances of places, are exposed. Like the practice turn within consumption studies, the performance turn dislocates attention from symbolic meanings and discourses to embodied, collaborative and technologized *doings* and *enactments*.

In contrast to the many studies portraying tourism as an overdetermined stage in which tourists are reduced to passive consumers that follow prescribed routes, the performance turn insists on uncovering creativity, detours and productive practices as much as choreographies and scripts. Much tourism research focuses upon how tourism companies and organizations, through guides, brochures and web pages, design destinations by inscribing them with place myths and staging them in postcard fashions (Larsen, 2006). And they ascribe great power to such symbolic design work in shaping tourist places and choreographing tourists' vision and cameras. Echoing many others, Urry uses the metaphor of the 'hermeneutic circle' to portray the choreographed nature of actual sight-seeing. He states that:

> much tourism involves a hermeneutic circle. What is sought for in a holiday is a set of photography images, which have already been seen in tour company brochures or on TV programmes. While the tourist is away, this then moves on to a tracking down and capturing of those images for oneself. And it ends up with travellers demonstrating that they really have been there by showing their version of the images that they had seen before they set off.
>
> (Urry, 2002a: 129)

Thus, effectively people travel in order to see and photograph what they have already consumed in image form. This model essentially portrays commercial imagery as all-powerful design machinery that turns the performances of tourists into a ritual of quotation whereby tourists are framed and fixed rather than framing and exploring (Osborne, 2000: 81). Many conceptual writings about tourism 'rapidly pacify the tourist – that is they tend to experience, perceive and receive but not do' (M. Crang, 1999: 238). A too fixed focus upon already inscribed destinations and staged experiences renders the tourist a passive sight-seer consuming sites in prescribed fashions.

We need a *circuit* of performance model that blurs the distinction between production (choreographing) and consumption (acting) and instead see them as interrelated and overlapping in complex ways. Tourist performances do not exist independently of structures of 'production' and wider societal discourses. Surely, as Edensor (2001a: 71) says, 'when tourists enter particular stages, they are usually informed by pre-existing discursive, practical, embodied norms which help to guide their performative orientations and achieve a working consensus about what to do'. Such 'norms' are guidelines, blueprints and nothing more (or less). 'Bodies are not only written upon but also write their own meanings and feelings upon space in a continual process of continual remaking' (Edensor, 2001b: 100). Performances are never determined by their choreographing, since there is always an element of unpredictability: the places and performances that tourists enact are never completely identical to the scripts in marketing material and guidebooks (Larsen, 2005). Portraying tourists solely as consumers disregards the fact that they *produce* photos and place myths; in the act of consuming, tourists turn themselves into producers. The act of 'consumption' is simultaneously one of production, of reinterpreting, re-forming, redoing, of decoding the encoded. Likewise, performances of production are partly shaped by, and respond to, performances of consumption (du Gay *et al.*, 1997).

Although tourism performances are surely influenced by guidebooks, concrete guidance, promotional information and existing place myths, the performance turn argues that tourists are not just written upon, they also enact and inscribe places with their own stories and follow their own paths.

In much tourism writing, places are presumed to be relatively fixed, given, passive and separate from those touring them. The performance turn destabilizes such static and fixed conceptions of places and sites. Instead places and performances are conceived as non-stable and contingent enactments. Edensor argues:

> The nature of the stage is dependent on the kinds of performance enacted upon it. For even carefully stage-managed spaces may be transformed by the presence of tourists who adhere to different norms. Thus stages can continually change, can expand and contract. For most stages are ambiguous, sites for different performances.
>
> (Edensor, 2001a: 64)

In this view, tourism performances are not separated from the places where they happen; they are not taking place in inert and fixed places. Tourist places are produced places, and tourists are co-producers of such places. They are performances of place that partly produce and transform places and connect them to other places. Most tourist places are 'dead' until actors take the stage and enact them: they become alive and transformed each time that new plays begin, face-to-face proximities are established and new objects are drawn in. Indeed, it can be argued that places emerge as tourist places, stages of tourism, only when they are performed (Bærenholdt *et al.*, 2004).

Thus, studies of tourist performances highlight how tourists not only consume experiences but also co-produce, co-design and co-exhibit them, once they enact them and retell or publish them afterwards. Publication has escalated with the significant rise of user-generated Web 2.0 sites such as social networking sites (e.g. Facebook, MySpace), photo communities (e.g. Flickr, Photobucket) and travel communities (e.g. VirtualTourist, TripAdvisor), where users produce web content as well as consuming it. And even when tourists do 'sight-see' they are not completely passive; most are busy making, for instance, photographs. The performance turn acknowledges that in the act of consuming tourists turn themselves into producers; they create, tell, exhibit and circulate tales and photographs that produce, reproduce and violate place myths that tourism organizations have designed and promoted.

Simultaneously, the analytical distinction between producer and consumer that has been so durable in tourism and marketing theory is dissolved. Furthermore, the performance turn has challenged representational and textual readings of tourism, in which bodies and places often end up being reduced to 'travelling eyes' by making *ethnographies* of what humans and institutions do – enact and stage – in order to make tourism and performances happen. So the performance turn represents a move to ethnographic research in tourism.

Finally, the performance turn does not see tourism as an isolated island but explores connections between tourism, the everyday and significant others such as family members and friends. Most tourism performances are performed collectively, and this sociality is in part what makes them pleasurable (Bærenholdt *et al.*, 2004). Tourism is not only a way of experiencing (new) places and events, but also entails emotional geographies of sociability; of being together with close friends and family members. And although there are elements of creativity and the unpredictable in tourism performances, they are also full of everyday conventions about proper behaviour and so on.

The performance turn has transformed tourism studies into an exciting and lively research field for those dissatisfied with purely representational accounts of tourism. We may say that the performance turn's major achievement is that it gives vibrant life, liveliness, happen-ness and joyful banality to tourist studies' former world of lifeless tourists, eventless events and dead geographies, and this book aims at following this path. And yet we also travel to and across new theoretical and methodological paths to overcome some of its limitations.

The reason for doing this is that we believe that the notion of performance has been incorporated a little too automatically and sketchily into tourist studies, sometimes solely because of its metaphorical appeal. There is, we think, a need to dwell a bit more theoretically and methodologically on the concept of performance. Not surprisingly, the performance turn draws heavily on Erving Goffman's dramaturgical sociology, in part because 'the cultural competencies and acquired skills that make up touristic culture themselves suggest a Goffmanesque world where all the world is indeed a stage' (Franklin and Crang, 2001: 17–18).

Goffman's notions of 'front-stage' and 'back-stage regions' are central to MacCannell's (1976/1999) now classic work on the staging of authenticity in modern tourism. 'The front is the meeting place of hosts and guests or customers and service persons, and the back is the place where members of the home team retire between performances to relax and to prepare' (MacCannell, 1976/1999: 92). MacCannell's argument was that tourists desire authenticity, or back-stage realities, but they often only encounter 'staged authenticity' put on show in front-stage regions. Likewise, Phil Crang (1994, 1997) utilizes Goffman's framework to discuss how tourism as a service economy is typified by face-to-face performances. He shows how waiting work in a diner-style restaurant is a form of conscious acting that is simultaneously scripted and creative taking place before the dining audience. Tim Edensor (1998, 2000, 2001a) also imports much of Goffman's language, speaking of tourists as improvising performers, actors and cast members, sites as stages, guides as directors, and so on (see also Erickson, 2004; Larsen, 2009).

One problem with a performance metaphor in social theory is that Western assumptions of performance are make-believe and illusion; outside the world of the 'theatre', performing is deception, a trickster world of false impressions. The common ontological distinction so prevalent in Western modernity between an authentic world of natural being and inauthentic one made up by performers has haunted tourist studies for a long time. For example, in MacCannell's (1976/1999) writing, it sometimes seems that modern tourism is nothing but performative illusions because it is a mobile world of *staged* authenticity: modern tourism is therefore permeated with inauthenticity. However, we will argue that all cultures and places are constructed through performances and connections with other places and therefore in a sense contrived or inauthentic; they are fabrications in the sense of something made (Rojek and Urry, 1997; Duncan and Gregory, 1999: 5). Moreover, Goffman's argument was not that all the world is a trickster stage. Goffman's dramaturgical sociality is embedded with a thick moral universe, and his claim is not that our social world is a theatrical one but only that a performance lens is a revealing optic on social life. Similarly, we argue not that tourism is a performance but only that a performance approach makes interesting studies of how tourism takes place and is practised (Larsen, 2009).

Although a Goffmanian lens is central to this book too (see especially Chapter 7), we want to argue for the need to develop the idea of performance and in part move beyond Goffman. In particular, we think it necessary to materialize the notion of performance. In what follows we briefly flesh out three significant contributions in this respect. First, we argue that James Gibson's (1977, 1979, 1982) notion of 'affordance' can help us to understand how performances are disabled and enabled by specific objects and material environments. Second, we relate the notion of affordance within what Nigel Thrift (2007) has coined 'non-representational theory' and argue that this approach may help us grasp the embodied, performative and material character of everyday life as well as the inescapable hybridity of 'human' and 'non-human' worlds. Finally, we turn to the discipline of performance par excellence, 'performance studies', in order to consider the performative character of everyday practices in more depth.

Materializing tourism performance

Material affordances

Despite the fact that tourists routinely engage with multiple material cultures, tourism studies has largely failed to understand the significance of materiality, objects and material networks for human performances. Studies of tourist sites and performances have generally been overly concerned with humans, thus neglecting the role of non-humans. In this book we emphasize how objects and technologies are crucial in tourism performances, primarily because they have a *use-value* that enhances the physicality of the body and enables it to do things and sense realities that would otherwise be beyond its capabilities.

The use of theatrical metaphors in social and cultural studies has broadly been accused of thinking of space as an empty stage, as a relatively fixed container for social actions that fails to consider how the spatiality of materials and other bodies is inscribed (and inscribes) with choreographies and performances by (and onto) the performing body. Inspired by Gibson we may ask how material spaces *afford* different performances.

Gibson's work was highly inspired by solving the problems of visual perception. Thus, his work has been highly influential in media, technology and design science (Norman, 1999). For J. J. Gibson (1982: 120), however, visual perception is dependent on the embodied movement through and involvement in a particular 'terrain, field or space'. What enables humans to judge distance, pace dimensions, act, re-act and move are precisely the material qualities of the environment, its contours, paths, texture, and so on, and the ways these are perceived, for example, from aircrafts and cars. Thus, actions and perceptions are based on profoundly embodied and technologised performances. For J. J. Gibson, materiality is not a 'neutral' backdrop to performances, nor is it a social and cultural construct. Material objects and surfaces afford particular performances to take place. Gibson's idea of affordance is closely related to an 'ecological' view on biological beings and their environments in which materialities (environments) and embodied beings (animals) are conceived of as complementary:

> The affordances of the environment . . . are what it *offers* to the animal, what it *provides* or *furnishes* either for good or ill.
>
> (J. J. Gibson, 1979: 129)

The practice of catching a ball, for example, depends on the material affordances of the ball and the environment: is it big or small, heavy or light, dry or slippery because of rain? All these qualities of the object and environment afford the catch to be performed in particular ways (with or without gloves, with one or two hands and so on). For Gibson, then, practices and their particular performances are always a blend of cultural norms and material properties. Whereas the use of the ball for playing baseball or football is, of course, culturally constrained and scripted by particular social scripts, the affordances of the ball (for being 'caught') is a property of the ball, the object itself. In Gibson's own words:

> The affordance of something does *not change* as the need of the observer changes. The observer may or may not perceive or attend to the performance, according to his needs, but the affordance, being invariant, is always there to be perceived. An affordance is not bestowed on an object by a need of an observer and his act of perceiving it. The object offers what it does because of what it is. To be sure we define *what it is* on terms of ecological physics instead of physical physics, and it therefore possesses meaning and value to begin with. But this is meaning and value of a new sort.
>
> (ibid.: 138–9)

Gibson has been criticized for harbouring a reductionist view on how technologies and tools tie into performances. However, his emphasis on how the artefacts, technologies and material environments enable and disable particular performances are not at odds with a view on performance as part of creative, improvisational everyday practices that are neither purely representational nor reducible to alleged underlying material or social layers. In line with Gibson (rather than against him), we can think of embodied everyday performances such as catching a ball, driving our car or touring a famous attraction as improvised performances afforded by the particular objects, technologies and environments we use, inhabit or 'dwell in'. It is only by such practical embodied day-to-day practices that the world of our everyday performances becomes a meaningful place. Like this, 'cultural' questions are reframed as a question also of the affordances of the environment. Inspired by Gibson, the anthropologist Tim Ingold – in a manner similar to non-representational theory, discussed below – reverses the relationship between performance and representation by insisting that 'meanings are not attached by the mind to objects in the world, rather these objects take their significance . . . they afford what they do, by virtue of their incorporation into a characteristic pattern of everyday activities' (Ingold, 2000a: 168).

Non-representational theory

Non-representational theory (Thrift, 1996, 2007) takes a modest ontological stance rooted in everyday *practices* that sees the 'social world' as continually (re)produced through performances of doing and acting. It:

> emphasises the flow of practice in everyday life as embodied, as caught up with and committed to the creation of affect, as contextual, and as technologized through language and objects . . . Clearly, then, a non-representational outlook depends upon understanding and working with everyday as a set of skills which are highly performative.
>
> (Thrift and Dewsbury, 2000: 415)

This approach valorizes *all* the senses, embodied skills and bodily enactments, and acknowledges that 'the "material" and the "social" intertwine and interact in all manner of combinations' (Thrift, 1996: 24). Being inspired by actor network theory, it sees bodies and technologies as intimately connected. Human performances are never purely 'social' or 'human', but tied up with tangible non-humans. They are thus hybrids.

Non-representational theory parallels a general turn towards theories of practice in consumption studies as well as a dissatisfaction with how practice has been conceived of, especially within the 'cultural turn'. Thus, Allan Latham (2003: 1901) remarks that 'Although the cultural turn pointed us towards events and processes marked by their mundaneness and ordinariness – in short their everydayness – its commitment to theorising and researching social practices has been patchy'. Especially in the UK, the cultural turn was dominated by a textual and representational understanding of social practices (less so in Scandinavia; see Simonsen, 1999, 2003), and the emergence of non-representational theory is partly a reaction against this. Briefly put, non-representational theory is concerned with bodily doings and technical enactments rather than with representations and meanings. It challenges the textual dominance, especially in cultural geography, by being concerned with ' "performative presentations", "showings" and "manifestations" rather than merely "representation" and "meaning" ' (Thrift, 1997: 127). Instead it is concerned with *practice* understood as an emergent outcome of embodied, precognitive and material performances. Non-representational studies are 'busy, empirical commitments to doings near-at-hand, in ordinary and professional settings, and through material encounters' (Lorimer, 2005: 84). The notion of non-representation may mislead, as representations also are understood as emergent outcomes of underlying performances. As Lorimer argues, we ought to speak of studies of the 'more-than-the-representational' (ibid.) as non-representational theory does not neglect representations but rather emphasizes that representations are produced and act in themselves. As formulated programmatically by a group of non-representational geographers:

> non-representational theory . . . is characterized by a firm belief in the actuality of representation. It does not approach representation as masks, gazes, reflections, veils, dreams, ideologies, as anything, in short, that is a covering which is laid over the ontic. Non-representational theory takes representation seriously; representation not as a code to be broken or as an illusion to be dispelled rather representations are apprehended as performative in themselves; as doings. The point here is to redirect attention from the *posited meaning* towards the *material composition and conduct* of representations.
>
> (Dewsbury *et al.*, 2002: 439)

The ambition is not to reveal a hidden truth or 'reality' beyond performance or representation, but to shed light on the performative aspects of 'the doings' of objects, artefacts, bodies, representations, and so on. In contrast to Butler's (1993) performativity theory of coded performances, Thrift uses the case of dance to make the argument that many social performances are pre-representational and pre-choreographed. Thrift argues that most embodied thinking and doing is non-cognitive, practical and habitual, but nonetheless potentially creative and non-predictable.

David Crouch's work on embodied tourism that crucially combines 'the material and the immaterial and the metaphorical' (Crouch, 2002: 210) has transported non-representational theory into tourist studies. Crouch's practice perspective celebrates the embodied nature of tourism and performativity as containing possibilities of the 'unexpected, the different, the risky' (Crouch, 2003a: 1946). He suggests that tourism essentially is a bodily practice. Tourists are 'surrounded by place' (Crouch, 2002) and their encounters are complexly sensual; they are embodied. Significantly, he argues that embodiment relates not only to physicality but also to imagination, fantasy, context and 'making sense' of doings. Tourists perform places sensuously, mentally and imaginatively; places exist on the ground, in mental landscapes as well as material cultures of images and objects (Crouch, 2002: 208). Moreover, Crouch is not opposing non-representational and representational thinking but can be said to be more-than-representational.

Performance studies

Although earlier primarily preoccupied with the performing arts, performance studies now embrace all sorts of ritualized and playful activities of the everyday. Like Goffman, performance studies claims not that all social life is a performance (as a play in a theatre is, by convention) but that most expressive everyday practices and behaviours – greetings, display of emotions, family unions, business meetings and so on – can be analysed *as* performance (the key here being 'as'). Richard Schechner (2006: 1) highlights four defining features of performance studies. First, the discipline's 'object of study' is actions and behaviour: 'what people do in the activity of their doing it'. It is concerned with the art of producing now. Second, studies of performances are understood as a performance too. Third, ethnographic participant observation is the privileged method. Fourth, few performance scholars favour detached observation; the norm is rather involvement. Schechner summarizes:

> Performance studies starts where most limited-domain disciplines end. A performance studies scholar examines texts, architecture, visual arts, or any other item or artifact of art or culture not in themselves, but as players in ongoing relationships, that is, 'as' performances . . . Briefly put, whatever is being studied is regarded as practice, events, and behaviors, not as 'objects' or 'things'. This quality of 'liveness' – even when dealing with media . . . – is the heart of performance studies.
>
> (Schechner, 2006: 2)

Later he argues that:

> To treat any object, work, or product 'as' performance . . . means to investigate what the object does, how it interacts with other objects or beings, and how it relates to other objects and being. Performances exist only as actions, interactions, and relationships.
>
> (ibid.: 30)

Clearly, non-representational theory and performance studies are closely related (for Thrift's sympathetic discussion of performance studies, see Thrift and Dewsbury, 2000). Both privilege practices over texts and study the 'liveliness' of social and technological life. They are concerned with actions. What typifies both approaches is that they deal with actions more than texts: with habits and expressive powers of the body more than structures of symbols, with the social construction of reality rather than its representation (Schieffelin, 1998: 194). And both acknowledge that non-humans play their part – as 'players', agents – in those networks that make performances possible. Technologies perform too: things do things, interact with other things and subjects, and enable things. As Schechner says:

If I were to treat the computer 'as performance', I would evaluate the speed of its processor, the clarity of its display, the usefulness of the pre-packed software, its size and portability, and so on.

(Schechner, 2006: 42)

For Schechner (2006: 30), the performances of a computer or any other technology or event depend not solely on its materiality, but also on its interactivity with other non-humans and technology, and such interactivity is always contingent and unpredictable. But performance studies is concerned not merely with doings (as non-representational theory seems to be), but also with what Schechner calls 'showing doings' or 'performative behaviour': 'how people play gender, heightening their constructed identity, performing slightly or radically different selves in different situations' (cited in Thrift, 2007: 113). Performance studies also pays more attention to the 'scripting' and rehearsed nature of performances than Thrift does. A key argument for performance studies is that performances always are 'restored behaviour': 'Performance in the restored behaviour sense means never for the first time, always for the second to the nth time: twice-behaved behavior' (Schechner, 2006: 37). Performances are never for the first time because they require rehearsal, imitation of other performances and adjustment to norms and expectations. And this is also the case with the performances of everyday life, when training and rehearsal revolve around learning appropriate culturally specific bites of behaviour, to such an extent that they appear natural, become taken-for-granted rituals. Individuals are never the single author of their performances; performances are always scripted by societal forces larger than the single individual, according to performance studies. Thus, performances are also about choreographies, something that Thrift tends to neglect. Thrift's pre-social conceptualization of dance is mistaken: dance is always twice-behaved behaviour mediated by teaching, scripting, cultural norms and watching audiences (see also Nash, 2000: 658). To speak of performances without taking account of their staging and scripting is naive. Yet this does not mean that performance studies sees performances as pre-formed *or* predictable: 'Performances can be generalized at the theoretical level of restoration of behaviour, but as embodied practices each and every performance is specific and different from every other' (Schechner, 2006: 36–7). This balance between 'restoration', or structure, and 'difference', or agency, can also be illustrated by two other key concepts in performance studies: ritual and play. To cite Henry Bial (2004: 135): 'In performance studies, play is understood as the force of uncertainty which counterbalances the structure provided by the ritual. Where ritual depends on repetition, play stresses innovation and creativity. Where ritual is predictable, play is contingent'. The point is that performances always have elements of both ritual and play. Hence, following Schechner, we can conceptualize performance as a form of playful ritualized behaviour: partly constrained, partly innovative. Against the tendency to highlight performances as pre-formed or subversive, even free, here 'agency, creativity, structure, and constraint become simultaneous', in Hughes-Freeland's words (1998: 8). Following Giddens's (1984) structuration theory of social practices, we may say that performance studies work with a duality (not a dualism) of structure (ritual) and agency (play). Such a performance outlook emphasizes human life as embodied, creative, interactive and tied up with enabling discourses and technologies. Performance studies thus argues that 'performance is located at the creative, *improvisatory* edge of practice in the moment it is carried out – though everything that comes across is not necessarily knowingly intended' (Schieffelin, 1998: 199).

Although Edensor does not discuss performance studies, his approach to tourism performances has much in common with this discipline. Edensor portrays performances of tourism as involving ambivalent relationships between home and away, tourists and the industry, possibilities and constraint, creativity and conducting. Thus, against the tendency to highlight performances as pre-formed or subversive, even free, Edensor stresses that performance 'can be conceived in more ambivalent and contradictory terms . . . as intentional and unintentional, concerned with both being and becoming, strategically and unreflexively embodied' (Edensor, 2001a: 78). When stepping into particular stages, pre-existing discursive, practical, embodied norms and concrete guiding by guides and signs usually choreograph tourists (ibid.: 71). Such choreographies are guidelines, blueprints, and nothing more (or less), and they enable as much as they constrain creativity. The tourist body 'is therefore both a physical entity with an immediate geography and culturally defined in terms of a style or body use' (Rodaway, 1994: 35). For that reason a performance perspective must always be connected to performativity. In the words of Nicky Gregson and Gillian Rose (2000: 434): 'Performance – what individual subjects do, say, "act-out" – and performativity – the citational practices which reproduce and/or subvert discourse and which enable and discipline subjects and their performances – are inherently connected, through the saturation of performers with power'. Tourist bodies are simultaneously pre-formed *and* performing, as we show throughout this book.

Mobilizing consumption and everyday life

Too often tourism has been approached as a segmented practice confined to particular spaces and times. Even though the performance turn has emphasized the everydayness of tourism performance, studies of tourists' performances tend to take place at emblematic tourist sites. In contrast, this book highlights how tourism performances happen across multiple sites, ordinary as well as extraordinary, where many mundane everyday practices are also performed. And although we travel along with tourists to the Orient we also make much shorter trips to their private homes to explore the afterlife of their tourist objects, images, memories and imaginations. A central argument of this book is that once the everyday is part of one's lens there is an urgent need to de-exoticize tourism theory.

Ethnographies of tourist performances have proved successful in analysing embodied tourism performances within a particular site, but unsuccessful in following flows of performances between and across sights/sites. Instead, such studies have largely neglected the networked mobilities of objects, images, texts and technologies that permit tourism performances to take place and be represented and (re)circulated across often great distances at various sites and times. In other words, one unintended consequence of bringing the performing tourist into focus is the neglect of the mobilities and networks of objects, information, communications and people that afford tourist performances, make tourist places and connect home and away.

Tourism is, we argue, not only a significant part of the ongoing transition of consumption within an emergent global cultural industry but also a prime driver of the *mobilization* of the everyday that supports this transition. In what follows we briefly moor this argument in some of the key theoretical discussions of contemporary changes in the structure of consumption and cultural production as well as the increased acknowledgement of mobility for constituting the social relations of society at the beginning of the twenty-first century.

Technological changes in transport and production as well as information and communication technologies (ICT) have transformed consumption. One such transformation is manifest in relation to the internet, as part of the much hyped, so-called Web 2.0 revolution, which affords an open online participatory culture in which connected individuals not only surf but make a lot of things too by editing, updating, blogging, remixing, posting, responding, sharing, exhibiting and tagging, and so on. In a broader perspective, the Web 2.0 highlights how consumers become part of the production process. As David Beer and Roger Burrows, in their article 'Sociology and, of and in Web 2.0: some initial considerations', say:

> Perhaps the key-defining feature of Web 2.0 is that users are involved in processes of production and consumption as they generate and browse online content, as they tag and blog, post and share. This has seen the 'consumer' taking an increasingly active role in the 'production' of commodities (Thrift 2005). Indeed, it is the mundane personal details posted on profiles, and the connections made with online 'friends', that become *the* commodities of Web 2.0. It is the profile, the informational archive of individuals' everyday lives, that draws people into the network and which encourages individuals to make 'friends'.
>
> (Beer and Burrows, 2007: 3.3)

Web 2.0 also has the potential to have much impact on tourism businesses and how tourists plan their journeys as tourists 'post' 'travel tales' not only to their significant others but also to 'strangers'. While it has become normal to book tickets and holidays on the internet, allowing for more individualized, post-Fordist travel patterns (see Lash and Urry, 1994), Web 2.0 gives tourists opportunities to publish their recommendations, reviews and photographs on various travel web sites such as www.tripadvisor.com and www.virtualtourist.com to other tourists so that they can plan their journeys *without* consulting the tourism industry's glossy brochures and home pages. Such travel community websites featuring user-contributed travel guides are likely to be conceived as more sincere than the always-glossy brochures and home pages of the tourism industry. These websites are 'powerful' in the sense that they can support or harm, be in line with or out of touch with, for instance, a city's official brand (see Ek *et al.*, 2008). Place branding is no longer only in the hands of the tourism industries; tourists are now part of that place-making process.

Another transformation of consumption is that cultural consumption practices today seem to be moving into a transitional phase from (mostly nationally) organized and regularized forms to more globally and unpredictable forms. This is particularly true for the cultural industries whose production, circulation and consumption are thoroughly transnational, mobile and fluid. Traditionally, literature on the cultural industries has conceived of the industrialized production of media and cultural products as the production of (alienated) symbolic meanings. Following Scott Lash and Celia Lury (2007), we may, however, argue that the twenty-first century has seen the emergence of a new mode of operation in the way in which cultural industries work. With the transition to global cultural industries:

> culture is so ubiquitous that it, as it were, seeps out of the superstructure and comes to infiltrate, and then take the infrastructure itself. It comes to dominate both the economy and experience in everyday life. Culture no longer works – in regard to resistance or domination – primarily as a superstructure. It no longer works as hegemonic ideology, as symbols, as representations. In our emergent age of global culture industries where culture starts to dominate both the economy and the everyday, culture, which was previously a question of representation, becomes *thing*ified. In classical cultural industry – in terms of both domination and resistance – mediation was primarily by means of representation. In global culture industry instead is the *mediation of things.*
>
> (Lash and Lury, 2007: 4)

What was previously a question of representation now becomes materialized, as objects, goods and products spin out of control and take on a life of their own in homes and cityscapes. Culture is now an all-pervading feature of our material environment and consumption. Even films, television productions and other media products play an important (but not dominant) part of the consumption of material places, artefacts and so on.

This means that the symbolic value of cultural products increasingly becomes unstable and unpredictable as objects (and we might add bodies, images and so on) *move.* Instead of approaching the use of cultural objects as simply the consumption of symbol, circulation may be viewed as a value-adding process. Again, according to Lash and Lury:

> products no longer circulate as identical objects, already fixed, static and discrete, determined by the intentions of their producers. Instead, cultural entities spin out of the control of their makers: in their circulation they move and change through transposition and translation, transformation and transmogrification. In this culture of circulation . . . cultural entities take on a dynamic of their own: *in this movement,* value is added.
>
> (ibid.: 4–5)

Lash and Lury do not ask whether the increased importance of circulation and mobilization seen in relation to cultural consumption signifies a more fundamental transformation of global society. Yet their focus on movement and circulation for the production of culture coincides with broader discussions in anthropology and cultural studies of how movement and circulation produce new forms of hybrid culture, belonging, identity and so on (Appadurai, 1999; Fortier, 2000). Most famously, James Clifford argues for tracing out the 'routes' of cultures and not only their 'roots' (Clifford, 1997). Not only objects move: imaginations, people and information also alternately mobilize and demobilize places and objects in cultural consumption. Thus, the production, circulation and consumption related to, for example, films such as the *The Beach* (2000) or *Lord of the Rings* (2001–3) rest on an underlying web of interdependent mobilities of not only objects, but also people, dreams, representations and information, as places, people and objects are mobilized, demobilized and reconfigured (Tzanelli, 2007).

This has also been an important theme for proponents of the 'mobility turn' in the social sciences. For Urry (2000, 2007) the mobility of people, images and objects is not only brute observable movements between territorial organized regions of the globe but an ontological condition transforming social relations from within. Thus, 'multiple and intersecting mobilities seem to produce a more "networked" pattern of social life' (Hannam *et al.*, 2006: 2). Mobilities transform the human condition on all scales, from the human body, and its increasingly prosthetic connectedness to other bodies through technological gadgets such as hands-free mobiles, to global processes of climate change. Urry (2007: 47) distinguishes between five different forms of mobilities: *corporeal* travel; the physical movement of *objects; imaginative* travel (the movement of places and people through different sorts of visual and print media); *virtual* travel (for example real-time presence across geographical distance mediated by wireless broadband connections); and *communicative* travel (through emails, text messages, mobile phones and so on). What is new, according to Urry, is not the presence of each of these mobile technologies, but the fact that, conditioned by their scale, ubiquity and convergence, they produce a new emergent social reality (ibid.: 195). So, rather than examining the scale and scope of particular forms of mobilities, what is interesting for social theory and studies is 'the complex assemblages between these five mobilities that may make and contingently maintain social connections across varied and multiple distances' (ibid.: 48). Importantly, Urry highlights that any study of any of these mobilities cannot be undertaken without careful attention to the necessary immobile material platforms – what he metaphorically calls 'moorings' – such as transmitters, roads, garages and airports that facilitate these mobilities. The key here is to work with a dialectical relationship between mobilities on the one hand and the immobilities and moorings on the other (Hannam *et al.*, 2006).

Tourist places and performances are continually being shaped, produced, reproduced and eroded as various:

Mobilities of people and objects, airplanes and suitcases, plants and animals, images and brands, data systems and satellites, all go into 'doing' tourism'. At the same time tourism also concerns the relational mobilizations of memories and performances, gendered and racialized bodies, emotions and atmospheres. Places have multiple contested meanings that often produce disruptions and disjunctures. Tourism mobilities involve complex combinations of movement and stillness, realities and fantasies, play and work.

(Sheller and Urry, 2004: 1)

In this way, the proliferation of tourist travel is a significant mediator of the *mobilization* of society. Lash and Lury (2007) and Urry (2007) point to the importance of conceiving contemporary society and culture through the lens of mobility. Following this, we attempt an analysis of tourism as an everyday performance that is not limited to the two weeks of holidaying or reducible to a desire for Otherness. Instead, we approach tourist performances as deeply embedded in the fabric of everyday life. In doing this we argue that tourism is an especially interesting field to examine how transformations in cultural consumption and social networks enable new patterns of consumption and forms of belonging in the world to emerge.

Performing tourism, performing the Orient

Studies of tourism and studies of the Orient are illustrative of a convergent evolution in social theory and science. Although arising from very different fields of study, respectively contemporary consumption and colonial domination in the seventeenth and eighteenth centuries, they have a range of similar traits. Both emphasize the production of difference as the key for explaining the preoccupation of the Western consuming Subject with the Other. Both depict Self and Other as mirror images reflecting distorted fragments of each other.

As Said (1995, originally 1978) argued three decades ago, the Orient was always a modern invention of the West. It became constructed as a projection, a fabrication, of the Occident's imagination, fuelled by 'his' desires, fantasies, wants and fears. Similarly, tourist theorists have most often pointed to the desire for difference as the logic behind modern tourist consumption, and it has been convincingly argued that Orientalism has delivered the key tropes that were generalized and translated into the leisurely consumption of difference. Thus, Sheller shows how Orientalist tropes migrated into the discursive construction of the Caribbean as an object for cultural consumption and tourist desires (Sheller, 2003: 109–10). Similarly, Hendry shows how the Orient strikes back by imitating scenery and landscapes of a number of (European and other) Others, and by doing this mimics the procedures and techniques of Orientalism (Hendry, 2001). In this way the tropes and techniques of Orientalism have been diligently imitated, used and recirculated by global cultural industries in general and in tourism consumption in particular. Both the desires and threats of the Orient have been frequently invoked, silenced and recirculated in continuously new forms.

In the course of working on this book we were constantly reminded of the complex interplay between tourism flows between the 'West' and the 'East' and the workings of what has been termed 'practical orientalism' (Haldrup *et al.*, 2006). Our research was carried out in Turkey and Egypt in 2005–2008. Thus, it took place in the shadow of the September 11 events, as well as terrorist attacks in tourist destinations in Egypt and Turkey that resulted in the occasional closure of these countries to West Europeans (and North Americans). And not least among the events occurring during this period was the so-called 'cartoon controversy' (see Chapter 5) induced by the publication and circulation in 2006 and 2008 of 12 cartoons of the Prophet Mohammed by newspapers in Denmark (where the two authors and most of the interviewees live). The cartoon crises caused consumer boycotts, demonstrations and riots aimed at Danish products, embassies and representatives throughout the Middle East (including Turkey and Egypt). Since then, however, both countries seem to have stabilized their position on the world tourist market as well as in relation to Danish mass tourism. As a result of the divergent mobilizations and demobilizations of Turkey and Egypt as 'tourist places' (or 'danger zones') during the research for this book, parts of the fieldwork had to be cancelled or rescheduled. However, these mobilizations and demobilizations also point to several interesting observations regarding the emergence of mobile society and everyday life. First, they show how tourists' imaginations are fragile and in constant danger from erosion due to pollution from 'non-tourist' networks and flows (such as those related to the Global War on Terror, separatism or political crisis). Second, they show how the interdependencies between different mobilities (the virtual and communicative mobility of the cartoons) and corporeal travel (tourism) provide unpredictable outcomes. Third, they show that the fragility and unpredictability associated with tourism have been integrated into people's everyday life. As 'the global' increasingly becomes part of the everyday, our mundane performances presume an acceptance of the unpredictability and vulnerability of travel and tourism. Terror-ridden destinations (such as those in this book) may fall out of the tourism circuit for shorter or longer periods of times; however, if they re-emerge they tend to do so quickly.

In this book we show how tourist travels to exotic places can no longer (if ever) be adequately captured through the binary model of the Western gaze consuming the repulsive or seductive Other. Instead, we have to trace out the many ways in which such travel connects to the may current mobilizations of everyday life induced by media, technology, transport, popular imagination, global warfare and so on.

In Chapter 2 we de-exoticize tourism theory and connect tourism and the everyday. We then, in Chapter 3, argue that in order to grasp the material mobilities mediating between tourist and everyday lives we have to 'follow the flows' between spaces of home and away. We point to the need to investigate tourism as consisting of multisited performances bridging those spaces rather than keeping them apart. The chapter coins the notion 'mobile tourism ethnographies'. In Chapter 4 we outline how tourist performances may be conceived of as an embodied and material practice – how the materialities of tourist spaces afford particular performances of tourist life, and tourist memories and emotions are thoroughly saturated with objects mediating and interconnecting home and away. Chapter 5 mobilizes 'the Orient', and suggests conceiving of Orientalism as established through intersecting everyday flows of objects, images and bodies between the West and the East. In Chapter 6 we empirically bring tourist studies out of the reserve by literally following in the footsteps of tourists from their arrival in the airport to their resorts, on sight-seeing trips, while playing by the pool, reading on the beach, and so on. We show how holidaying is structured through particular mobilities, immobilities and moorings. And we explore the scenes for tourist practices and show how performances of home and Other complement each other in tourists' everyday lives on holiday. In Chapter 7 we then focus on one emblematic tourist practice, photography, and scrutinize ethnographically how the new affordance of digital photography is performed by photographing tourists and changing the nature of tourist photography. After having ethnographically traced out the detailed performances of tourists on holiday we follow the flows of images, objects and bodies back home to examine the afterlife of tourism in Chapter 8. We visit people in their private homes and explore how souvenirs (including personal photographs) are stored and displayed in the home as well as distributed beyond the home. Finally, in Chapter 9, we discuss how people spatialize their everyday life within social and cultural geographies of travel and tourism. We show how tourist travel feeds into the production and reproduction of different modes of being in the world and weaving relationships to people, places and cultures.

2   De-exoticizing tourist travel

Introduction

In tourist studies, and the social sciences more broadly, tourism is often treated as an exotic set of specialized consumer products occurring at specific times and places which are designed, regulated or preserved more or less specifically for tourism, such as resorts, attractions and beaches. Much tourism theory, such as the work of MacCannell (1976/1999) and Urry (1990, 2002a), defines tourism by contrasting it to home geographies and 'everydayness': tourism is what they are *not*. It is 'a no-work, no-care, no-thrift situation', according to E. Cohen (1979: 181). Or, to cite Suvantola:

> Travel can be conceptualized as a move from the mundane everyday routines of the home with its routines and sameness, to experience the excitement of being away from home and experiencing the Other. Travel can be thought of as a dialectic which involves leaving home, being in contact with the Other and then returning home.
>
> (Suvantola, 2002: 81)

In *The Tourist Gaze* (1990), Urry argues that tourism is formed in opposition to everyday life: a 'key feature would seem to be that there is a difference between one's normal place of residence/work and the object of the tourist gaze . . . Tourism results from a basic binary division between the ordinary/everyday and the extraordinary' (p. 11). In this early work, the distinction between home and away, ordinary and extraordinary, is the identifying regulator of what may come to be constituted as an extraordinary place of the tourist gaze. The main focus of such research is on the *extra*ordinary, on places elsewhere. Tourism is an escape from home, a quest for more desirable and fulfilling places. As a result, tourism studies too often produce fixed dualisms between the life of tourism and everyday life – extraordinary and ordinary, pleasure and boredom, liminality and rules, exotic others and significant others, to mention some. Such 'purification' means that everyday life and tourism end up belonging to different ontological worlds, the worlds of the mundane and the exotic respectively (exceptions are Crouch, 1999; Löfgren, 1999; D. Williams and Kaltenborn, 1999; McCabe, 2002; Franklin, 2003; Hall, 2005; Uriely, 2005; Edensor, 2006).

This chapter further develops the performance turn (see Chapter 1) within tourism studies by discussing the problems of this separation between tourist travel and everyday life: tourist geographies and home geographies. It takes inspiration from the everyday life theorist Lefebvre's (1991) claim that all aspects of social life are infused with elements of everyday life: no practices escape 'everydayness'. The chapter conceptually discusses some of the ways in which everyday life permeates tourism consumption, and especially how 'tourist escapes' are informed by everyday performances, social obligations and significant others. Discussion of everyday life is mainly absent from tourism theory and research: it is merely tourism theory's mysterious Other: everywhere and nowhere, known and yet unknown. This chapter therefore begins with a theoretical discussion of everyday life. It is argued that one significant aspect of everyday life is routine, and that this is part of the traveller's baggage. However, it is simultaneously stressed how everyday performances have potentials for creativity and the unexpected, and how many everyday spaces are sites of tourist consumption. Then the significance of significant others and face-to-face sociality in relation to everyday life are discussed. The main part of the chapter is concerned with reviewing, elaborating upon and bringing together various research projects within leisure studies, tourism studies, migration studies and social networks analysis, addressing connections and overlaps between everyday life and leisure travel. Particular attention is paid to the influential work of John Urry (1990, 2000, 2002a,b, 2003; Larsen *et al.*, 2006, 2007) because it partly illustrates how leisure travel is no longer merely an escape from everyday life but also a way of performing it. Although Urry's hegemonic concept of the tourist gaze was originally constructed on the premise that its opposite is ordinary everyday life, he later modified this by pointing out that the tourist gaze both is constructed *and* takes place through everyday media cultures. The emphasis now is on how post-modern media cultures saturate everyday life, which therefore itself becomes not merely grey and ordinary but also full of exotic signs and consumer goods. In this chapter we show how the performance turn destabilizes the tourist gaze and highlights how many tourist practices are embodied and habitual and involve *ordinary* objects, places and practices. We demonstrate how research on migration, diasporas, tourism and social networks shows how leisure travel is concerned with visiting and hosting significant others and attending 'obligatory' social events. We end by discussing research that shows how the everyday informs various spaces and performances among tourists as well as introducing the notions of cosmopolitan and national home-making.

Spatialities and mobilities of everyday life

It is to some extent understandable why tourism researchers have distinguished tourist travel from everyday life. In most of the everyday life literature, 'everydayness' is characterized by repetition, habitual practices, obligations and reproduction. As Edensor says:

The everyday can partly be captured by unreflexive habit, inscribed on the body, a normative unquestioned way of being in the world . . . The repetition of daily, weekly and annual routines . . . how and when to eat, wash, move, work and play, constitutes a realm of 'commonsense' . . . These shared habits strengthen affective and cognitive links, constitute a habitus consisting of acquired skills which minimize unnecessary reflection every time a decision is required.

(Edensor, 2001a: 61)

For example, Featherstone (1992: 165) contrasts everyday life with 'heroic life': 'The heroic life is the sphere of danger, violence and the courting of risk whereas everyday life is the sphere of women, reproduction and care'. Whereas 'heroic life' is male, unpredictable and nomadic, everyday life is fixed to a female and routinized domestic sphere. In this light, as travel has long been associated with masculine values of adventure and self-realization, travel seems to epitomize 'heroic life'.

However, this crude account of both travel and everyday life can be challenged. As Edensor points out, there is more to everyday life than the habitual, pre-scripted and ordinary. The classical texts of Lefebvre (1991) and de Certeau (1984) show the potential of everyday practices for creativity, subversion and resistance. In particular, de Certeau stressed the need for examining the 'tactics' that people in their everyday life employ to manipulate officially inscribed signs, objects and places:

the presence and circulation of a representation (taught by preachers, educators and popularizers as the key to socioeconomic advancement) tell us nothing about what it is for its users. We must first analyze its manipulation by users who are not its makers.

(de Certeau, 1984: xi)

In de Certeau's work, the everyday is the heroic realm of modernity, full of creativity, manipulation and resistance. As discussed by Hingham (2002), most writers on everyday performances highlight ambivalent relationships between possibilities and constraint, scripting and creativity, which reflect that everyday life is a complex notion. What is less discussed in the literature is the significance of sociality and social relations to everyday life. The classic texts of Simmel (1950, 1997a,b) and Goffman (1959) are exceptions here. Both argue that humans are social beings and that most everyday practices are social interactions which take place in close (visual) proximity to other people. One major aspect of Simmel's work explores how modern cities create new experiences of proximity: 'modern times for Simmel are experienced largely through changing relations of proximity and distance and, more broadly, through cultures of movement and mobility' (Allen, 2000: 55). Simmel (1950) argued that people in the modern metropolis increasingly found themselves amongst strangers and they therefore had to learn the social skill of distancing themselves from the mobile crowd. Simmel adopted the figure of the stranger to illustrate the modern metropolis's unique geographies of proximity and distance: here people are close in a spatial sense, yet remote in a social sense. Yet Simmel also discusses everyday interaction among significant others. Simmel speaks of 'sociability' to denote those kinds of interaction characterized by free play and non-instrumental and emotional sociality. Sociability is a 'pure interaction' between, in theory, equal participants that come together for the sole purpose of enjoying each other's company. One example of sociability is the communal meal (Simmel, 1997b). More broadly, visiting and hosting friends and relatives – crucial leisure activities that often involve some travel – can be seen as emblematic forms of sociability.

Goffman's classic *Presentation of Self in Everyday Life* (1959) outlines a dramaturgical framework to describe everyday social interactions, amongst strangers in public places. For Goffman, the self is a performed character, a public performer with carefully managed impressions. People, as everyday actors, are reflexive and strategic agents moving between different socio-spatial stages (or regions) requiring and allowing specific performances. These are front-stages and back-stages. A public performance is put on show in front-stages; in back-stage regions these performances may 'knowingly contradict'; 'back-stages' allow masks to be lifted temporarily (Goffman, 1959: 114). Everyday life is described as fundamentally performative and put on stage for an audience. It is performed in the ambivalent space between pre-fixed choreographies and improvisational performances. Thus, performances are culturally scripted but not predetermined (for a more detailed discussion of Goffman's notion of performance in relation to photography, see Chapter 7).

Goffman's work illustrates how everyday life is performed in various places, but home is traditionally regarded the base (especially for women) for everyday life, the 'back-stage' where families can be themselves. Heller (1984: 239) writes:

> integral to the average everyday life is awareness of a fixed point in space, a firm position from which we 'proceed' (whether every day or over larger periods of time) and to which we return in due course. This firm position is what we call 'home'.

In contrast, geographers and transport researchers often consider everyday activities and travel to take place predominantly within a territorial bounded 'activity space' (Hägerstrand, 1985; Massey, 1995; Holloway and Hubbard, 2001; Ellegaard and Vilhemson, 2004).

We think that such an equation of everyday life with a physical home, face-to-face interaction and local activities is too static and a-mobile. As we discuss in Chapter 3, there is much evidence that everyday life, for many, is in part networked, mediated and distanciated. The home has become a communication hub infused with mobile messages and connected screens. The 'time–space compression' (Harvey, 1989) that such technologies create means that distant places travel in and out of our living rooms:

But most of us are on the move even if physically, bodily, we stay put. When, as is our habit, we are glued to our chairs and zap the cable or satellite channels on and off the TV screen – jumping in and out of foreign spaces with a speed much beyond the capacity of supersonic jets and cosmic rockets, but nowhere staying long enough to be more than visitors, to feel chez soi.

(Bauman, 1998: 77)

So, far from being grey and ordinary, our everyday spaces are full of exotic and spectacular signs; we may say that the tourist gaze has become part of our everyday life because we spend so much time in front of television and computer screens: screens through which the globe enters our (mobile) 'homes'. Consequently, in 1994, Lash and Urry proclaimed the 'end of tourism':

People are tourists most of the time, whether they are literally mobile or only experience simulated mobility through the incredible fluidity of multiple signs and electronic images.

(Lash and Urry, 1994: 259)

The tourist gaze is no longer set apart from everyday life, as it used to be in modern times, but has become part of it. There is a de-differentiation between tourism and everyday life (see also Rojek, 1993). Thus 'the end of tourism' actually means not less but *more* touristic gazing because 'the post-tourist does not have to leave his or her house in order to see many of the typical objects of the gaze' (Urry, 2002a: 90). Although Urry occasionally suggests that 'imaginative travel' through media cultures replaces 'corporeal travel', the 'end of tourism' thesis really suggests the 'touristification of everyday life' and de-differentiation between tourism, everyday life and various form of travel:

there is no evidence that virtual and imaginative mobility is replacing corporeal travel, but there are complex intersections between these different modes of travelling that are increasingly de-differentiated from one another.

(ibid.: 141)

Whereas Urry traditionally contrasted the tourist gaze with the everyday, we have seen how in his later work he argued that the tourist gaze mediated by and because of global media signs blurs home and away by 'exoticizing' home geographies and familiarizing faraway places, producing a condition of 'banal globalization' or a compressed global village in which we all become 'banal cosmopolitans' increasingly 'consuming the world from afar' (Szerszynski and Urry, 2006).

'Time–space compression' also seems to involve 'time–space distanciation' (Giddens, 1990), i.e. the geographical spreading of people's social networks. This is partly the result of recent increases in travel and in longer-distance communications through cheaper international calls, text messages and free emails (Wellman, 2002; Urry, 2003). Larsen *et al.* (2006) show how it has become common to have strong ties at a distance and sustain them through phone calls, text messages, emails and occasional visits. Socializing at a distance has become a significant everyday practice. The social sciences can no longer equate closeness and communion with geographical nearness and daily or weekly co-present visits (ibid.) and neglect the significance of long-distance travel, occasional sociality and mediated communication to the spatialities of everyday life. Most social life on weekdays revolves around localized areas, and routinized, brief trips, with many people undertaking longer journeys at weekends and during holidays (Axhausen *et al.*, 2002). Thus, everyday life research should analyse how everyday practices of caring and socializing *also* take place at a distance and how people increasingly need to travel to socialize with their significant others.

Although there is a physical and static element to home, it is paramount for the mobilities paradigm also to detect how home can be mobilized and connected to other places. Following Berger (1984), we can understand home not solely as being rooted in one particular physical place, but also as something that involves, and can be mobilized through, social habits, small daily rituals, precious objects, mundane technologies and significant others.

Thus, the notion of the everyday is complex. Some use the notion to highlight the quotidian whereas others speak of creativity and subversion. This complexity makes 'everydayness' a useful concept in relation to studies of tourist performances. On the one hand, it allows an analysis of how 'tourist escapes' are full of everyday practices such as eating, drinking, sleeping, brushing teeth, changing nappies, reading bedtime stories and having sex with one's partner as well as co-travelling mundane objects such as mobile phones, cameras, food, clothes and medicine (Duncan and Lambert 2003; Obrador-Pons, 2003). 'Even when a traveler leaves home, home does not leave the traveler' (Molz, 2006a: 340). Our approach is inspired by Molz, who argues that:

> home is not just a place, but also a process of regular patterns and social connections that may be performed and reiterated even while travelling . . . Rather than becoming impossible in the midst of movement, home continues to matter as a physical and emotional site of belonging.
>
> (Molz, 2008: 6)

And she continues: 'home in the world is very much embodied and embedded in everyday practices while on the road. Travelers enact home at this global register precisely through small, localized acts of habitability' (ibid.: 13). Home is therefore part of tourists' baggage and bodily performances. On the other hand, while neglecting the everydayness of everyday life (Felski, 1999), de Certeau's 'resistance' perspective can help to write more dynamic and open accounts of performances than is common in the tourism literature, in which tourists so often drown in a sea of signs and choreographies (Larsen, 2005). Following Simmel and Goffman, an everyday perspective also enables studies of the significance of significant others, sociability and role-playing to the tourism experience, to insert the social into tourism research. Tourism studies has mostly neglected issues of sociality and co-presence and thereby overlooked how much tourist travel is concerned with (re)producing social relations. We now discuss various tourism publications that explore connections between tourism and the everyday. We begin with Edensor's work on tourism performances.

Mundane and collective tourism performances

The performance turn *explicitly* conceptualizes tourism as intricately tied up with everyday practices, ordinary places and significant others, such as family members and friends, both co-residing and at a distance. Furthermore, we argued, following Edensor (1998, 2000, 2001a), that performances should be seen as potentially creative, unreflexive, unintentional and habitual. This differs from ideas of tourism as a liminal zone, in which everyday conventions are suspended (e.g. Shields, 1991):

> Rather than transcending the mundane, most forms of tourism are fashioned by culturally coded escape attempts. Moreover although suffused with notions of escape from normativity, tourists carry quotidian habits and responses with them: they are part of their baggage.
>
> (Edensor, 2001a: 61)

Elsewhere he argues:

> Many tourist endeavours are mundane and informed by an unreflexive sensual awareness, and hence not particularly dissimilar to everyday habits and routines.
>
> (Edensor, 2006: 26)

Tourists never just travel *to* places: their mindsets, routines and social relations travel *with* them. The imaginative geographies of tourism are as much about 'home' as faraway places. Such a focus upon everyday practices and the ordinary is particularly stressed by Obrador-Pons (2003) in his Heidegger-inspired dwelling perspective. Contrary to the notion that tourism performances mainly engage the visual sense, that they are extraordinary and somewhat aloof and disembodied, Obrador-Pons argues that tourism is a multisensuous and practical way through which we are involved in the symbolic and not least physical world; it is a particular way of being-in-the-world, of dwelling in it (this is further developed in Chapter 4). He uses the notion of dwelling 'because it enables a genuinely geographical and social account of tourism that prioritizes *everyday* embodied practices' (ibid.: 47, our italics). Tourism, he argues, is essentially about practising space and practising through space; it is about embodied *doings*:

> It is because we are doing something in a particular way that we are tourists and we adopt tourist consciousness. The most relevant embodied practices through which we become tourists are everyday ordinary, and often

non-representational, practices. It is, therefore, insufficient in tourist studies to focus only on extraordinary practices, like sight-seeing.

(ibid.: 52)

Obrador-Pons asks us to explore the many more or less 'ordinary' practices and places that are made pleasurable on a holiday through creative inversions and how tourists make themselves at home in foreign places (see H. Andrews, 2005). This requires that we de-exoticize tourism theory and adopt a non-elitist approach to tourism practices.

Dwelling and building is intimately connected in Heidegger's (1993) thinking. Leisure and tourism research has shown how allotments and summer cottages are significant places of working and dwelling where 'people are working intensively most of the time' through free creative play (Jarlöv, 1999: 231). Partly for that reason, summer cottages are places where people often put down a rooted sense of belonging (Jarlöv 1999; Löfgren 1999; D. Williams and Kaltenborn, 1999; Hall and Müller, 2004). But tourists in rented summer cottages can also be said to be dwelling. In *Performing Tourist Places* (Bærenholdt *et al.*, 2004), we explored how second-home tourism often contains strong elements of such a sense of dwelling. This came out strongly in diaries written by cottage dwellers:

> Up for an early morning bath, at the beach all day, bathing, building castles in the sand, collecting mussels at the beach, the children tumbling around in the sand, had lunch on the beach. Walked to our house, decorated the house with shells and stones, played cards with the children.
>
> (cited in Haldrup, 2004: 444; see also Bærenholdt *et al.*, 2004)

Heidegger's (1993) equation of building and dwelling is evident here. The family is domesticating the place of holidaying by building sandcastles and decorating the rented house with the collected shells and stone. By doing this they incorporate the material affordances of their vacation space in their mundane routines creating a place for dwelling.

Both Bærenholdt *et al.* (2004) and Löfgren (1999) bring out some of the social and emotional significance of ordinary tourist practices and co-travelling significant others to the tourism experience. Tourism studies have overlooked the fact that many tourists experience the world not through a solitary 'romantic gaze' or the 'collective gaze' of mass tourism (Urry 1990, 2002a), but in the company of friends, family members and partners. *Performing Tourist Places* opens with a *private* photograph of two families posing with spades and buckets on a beach in front of the sandcastle they have just built. The communal projects of building a sandcastle and taking photographs show how tourists not only bring their own bodies but travel and perform with other bodies too. Most tourism performances are performed collectively, and this sociality is in part what makes them pleasurable. These books demonstrate how tourism is not only a way of practising or consuming (new) places, but also an emotional geography of sociability, of being together with close friends and family members from home. While travelling together, couples, families and friends are actually together, not separated by work, institutions, homework, leisure activities and geographical distances. They are in a sense most at 'home' when away from the home. *Performing Tourist Places* speaks of 'inhabiting tourism' whereas Löfgren (1999) speaks of 'Robinsonian tourism'. Both concepts highlight how much tourism is bound up with performing social life and building an alternative 'home', a utopian performance in which everyday routines, doings and roles hopefully become extraordinary: relaxed, jointed and joyful. Tourists are not only questing authentic places and objects, they also seek authentic sociability between *themselves* (Wang, 1999: 364). 'Getting away from it all might be an attempt to get it all back together again' (Löfgren, 1999: 269).

By highlighting the collaborative nature of tourism performances, we bring forth how sociability (Simmel's notion) is much more central to tourism experiences than previously thought. Ethnographic studies also show how much tourist travel even to typical tourist sites is about sociability. Kyle and Chick's (2004) ethnography of an American fair and Caletrio's (2003) study of Spanish tourists on the Costa Blanca demonstrate how families repeatedly return to these places because they have turned into meeting places where they meet up with relatives and friends living elsewhere. Similarly, Jacobsen (2003), in his discussion of northern package tourism to the mass tourism destinations in the Mediterranean, argues that:

> Many vacationers travelling from one European country to another should instead be compared with those who go to their summer cottages or second homes within the nation, habitually living a relaxed (family) life for a few weeks. Most Europeans go abroad as couples, families and circles of friends, which implies that they unavoidably bring with them parts of their everyday horizon into novel and temporary holiday environs.
>
> (ibid.: 75–6)

According to B. Brown (2007), in his study of 'practical problems in tourism experiences' (the collaborative *work* that tourists need to invest to perform tourism properly, such as pre-trip planning and using maps), tourism is a source of sociability:

> arguing for the collaborative nature of tourism highlights its sociable (Simmel 1949) nature . . . [the collaborative work of tourists] can also act as a setting for interactions that are of value and enjoyment to tourists, simply because of the time spent in the company of others or those met while travelling.
>
> (ibid.: 379)

This focus on 'significant others' is also central to the new literature examining how some people travel to 'distant' places to *meet* significant others rather than merely 'consuming' the 'Other'.

Mobilities and meetings at a distance

As D. Williams and Kaltenborn (1999: 214) say: 'When we think of tourism we often think of travel to exotic destinations, but modernization has also dispersed and extended our network of relatives, friends and acquaintances'. And statistical data show that such extended social networks now generate much tourist travel. According to the World Tourism Organization (WTO), in 2001, 154 million international trips were for the purpose of 'VFR [visiting friends and relatives] health, religion, other', compared with 74 million in 1990 (http://www.world-tourism. org/facts/trends/purpose.htm). Although holiday visits to the UK are declining, more and more people are travelling to the UK to visit friends and family members (Travel Trends, 2004), and almost half of all long-distance journeys in the UK are made to visit family and friends (Dateline, 2003: 17, 57).

Clifford's (1997) notions of 'dwelling-in-travel' and 'travelling-in-dwelling' deconstruct distinctions between home and away by pointing to the possibilities of being at home while travelling and coming home and dwelling through travel. Now that travel and displacement are widespread, we need to rethink dwelling so that it is no longer the antithesis to travel or simply the ground from which travel departs and returns (Clifford, 1997: 44; see also Franklin and Crang, 2001: 6). In his *Sociology beyond Societies*, Urry (2000: 157) argues that there are

> a variety of ways of dwelling, but that once we move beyond that of land, almost all involve complex relationships between belongingness *and* travelling, within and beyond the boundaries of national societies. People can indeed be said to dwell in various mobilities.

Tourism and migration researchers are beginning to examine how 'tourism visits' are essential to the lives of migrants and individuals from diasporic cultures, who often have strong ties in multiple places and feel at home in more than one place. Migration is far from being a one-way journey, leaving one's homeland behind, and is often a two-way journey between two sets of 'homes' (Duval 2004a,b; Mason 2004), and this generates tourist travel. According to M. A. Williams and Hall (2000: 7; see also M. A. Williams *et al.,* 2000; Gustafson, 2002; O'Reilly, 2003; Coles and Timothy, 2004):

> Many forms of migration generate tourism flows, in particular through the geographical extension of friendship and kinship networks. Migrants may become poles of tourist flows, while they themselves become tourists in returning to visit friends and relations in their areas of origin.

Whereas diasporas and displaced people traditionally demonstrate a desire for a permanent return, today's migrants can fulfil their 'compulsion to proximity' (Boden and Molotch, 1994), the desire to be physically co-present with people and with their homeland, through frequent virtual and imaginative travel, and especially through occasional visits. Various studies show how many immigrants and their (grand)children regularly visit their 'homeland' and other displaced family members across the world to keep their 'national' belonging and family networks 'alive' (Kang and Page, 2000; Miller and Slater, 2000; Mason, 2004; Sutton, 2004). Duval (2004a,b) and Nazia and Holden (2006) illustrate how parents of Caribbean and Pakistani origin feel obliged to travel to their homeland and personally introduce its key features to their children. Social obligations to travel are often intricately intertwined with obligations to visit specific monuments and religious sites. Nazia and Holden (2006) call this 'the myth of return'. In 'Heimat tourism in the countryside: paradoxical sojourns to self and place', Soile Veijola (2006) fascinatingly addresses issues of home and belonging in a context of increasing mobility. She describes in autobiographical fashion how she has come to feel like a tourist or a guest in her own home town, and thereby brings out how distinctions between home and away, tourists and guests, are increasingly blurred.

This point was also stressed in a recent research project on social networks and travel carried out by Urry (and one of the authors) (Larsen *et al.*, 2006). In investigating the geographical 'stretching out' of social networks and its implications for sociality and travel among youngish architects, diverse employees in fitness centres and security staff living in north-west England, they found that it has become common to have 'strong ties' at a distance and to undertake regular long-distance travel to meet friends and family members. This is because of both the historically high levels of migration for work and education as well as the emergence of low-price long-distance travel and communication. Their respondents reported that, on average, they live some 400 km from their identified 'strong ties' and make 10 long-distance national journeys yearly, mainly to visit kin and friends, as well as to attend birthday and Christmas parties, weddings, stag or hen nights, and so on. They compensate for the intermittent nature of meetings and the cost of transport (time, money and weariness) by spending a whole day or weekend or even week(s) together. Research suggests that people are socializing with each other less frequently on a weekly basis, partly because networks are now more dispersed (see McGlone *et al.,* 1999; Putnam, 2001), but, according to Larsen *et al.* (2006), this is to some extent counteracted by travel at weekends and during holiday periods.

This research indicates that 'visiting friends and relatives' (VFR tourism) is desirable and indeed necessary because even highly regular phone calls, text messages and emails are not enough to reproduce strong ties, which also depend on periodic face-to-face meetings. Larsen *et al.* (2006) argue that the increase in 'VFR tourism' stems from a 'compulsion to proximity' and from various *obligations* that require physical co-presence. They point out that most tourism theories fail to notice the *obligations* that choreograph 'tourism escapes' and leisure travel more broadly (see also Urry, 2002b) and show that there are obligations that require face-to-face co-presence, such as birthdays, Christmas parties, funerals, hen nights, stag nights and the weddings of close friends and family members, even if they necessitate considerable travel. Failure to fulfil such social obligations often has significant social consequences: social faces and relationships are likely to be damaged. When social networks were socially and spatially close-knit, relatively little travel was required to meet social obligations; today, however, when social networks are widely distributed and the world is becoming compressed as a result of relatively cheap and fast transport, such obligations trigger much long-distance travel. Larsen *et al.* (2006) concluded that travel that would have once been classified as 'touristic', and by implication a matter of 'choice', seems to have become central to many people. Their findings are in line with Franklin and Crang (2001: 7), who argue that: 'Tourism has broken away from its beginnings and ephemeral ritual modern national life to become a significant modality through which [national and] transnational modern life is organised'.

Much of the research discussed in this section suggests that networking is now an illuminating concept to work with. As discussed elsewhere, networking highlights how leisure/tourist travel is a social practice that involves significant others, face-to-face proximity and non-commercialized hospitality. It further highlights how tourist travel is not only a way of seeing the world but also a way of socializing with significant others and attending obligatory social events. It suggests that the analysis of everyday practices, social obligations, networks at a distance and social capital should be central to twenty-first century leisure and tourism theory (Larsen *et al.*, 2007).

This also means that the term 'tourist' needs some deconstruction as people undertake leisure travel for many different reasons (see also Rojek and Urry, 1997). Those spending the summer in a second home or visiting their best friend who now happens to live abroad are not likely to consider themselves tourists in the same way as someone spending two or three weeks on a package tour. Indeed, there is the probability that they will not consider themselves tourists at all. Nonetheless, the WTO will include them as tourists in its statistics. The WTO uses overnight stays to differentiate between day trips and tourism, and leisure travel becomes 'tourism' whenever it involves an overnight stay, no matter where this takes place (hotel or private accommodation).

This definition is both problematic and constructive. It is problematic because it mixes forms of leisure travel that have little in common except perhaps the journey and the overnight stay. Moreover, it neglects that tourist consumption can take place at home. It is constructive because it highlights how tourism increasingly overlaps with other forms of mobility and has become central to much social life in ever more mobile societies. In the process of de-exoticizing theory, this chapter suggests depurifying the disciplines concerned with travel and mobility, such as leisure studies, tourism studies, migration studies and transport studies. Rather than a distinct discipline of tourism studies, therefore, we need to develop transdisciplinary mobilities studies (Urry, 2000; Coles *et al.*, 2005).

Familial places

Although we often think of tourism as taking place in extraordinary and unknown places, many tourism places are in fact familial places, i.e. places where tourists feel at home away from home. Various tourism scholars have invented terms such as 'environmental bubble' (E. Cohen, 1972: 166), 'tourist bubble' (V. L. Smith,

1977: 6), 'enclave of familiarity' (Farrell, 1979) and, the latest, 'enclavic tourist space' (Edensor, 1998) to portray the familial and home-like qualities of international hotel chains, mass tourism resorts and destinations more generally. In his writing on tourism performances in India, Edensor makes an analytical distinction between 'enclavic' and 'heterogeneous' tourism space to highlight how tourism performances occur on very different scenes, materially and symbolically. He uses the notion of enclavic space in relation to organized mass tourism; they are 'tourist bubbles' – resorts – that are clearly demarcated from *local* spaces, effectively shielded from their potentially offensive sights, sounds and smells and strange food. It is contrasted with the fluid heterogeneous space of *individualized* backpacker tourism, during which tourists and locals supposedly rub shoulders. Enclavic tourist spaces have a 'familial' feeling because, in them, tourists are surrounded by like-minded tourists, international interior and amenities, Western-style food, English-speaking staff, and so on. There is a clear colonial touch to such enclavic spaces in the developing countries that often happen to be former colonies such as India.

The familiarizing or de-exoticizing element to Mediterranean mass tourism, beginning in the late 1960s, is often one of 'travelling parochialism' (Jacobsen, 2003), or what Billig (1997) terms 'banal nationalism'. This is particularly evident in relation to food and drinks, as nationally themed bars and restaurants easily outnumber those presenting themselves as locally authentic; thus, tourists can eat and drink foodstuffs from their native land surrounded by tourists of the same nationality as well as by flags and other national symbols of their homeland (Jacobsen, 2003). Based upon a recent ethnography of Britishness in charter tourist resorts (Palma Nova and Magaluf) in Mallorca, H. Andrews (2005: 252; see also West, 2006) shows how:

> For example, place names are resonant of those found in the UK, with café-bars called *The Britannia*, *The Willows*, *The Red Lion*, and others that make reference to British popular culture – Benny Hill and Eastenders, for instance. Added to this, English is the main language spoken and British sporting fixtures, news and other television programmes are beamed in by satellite or played from video recordings. Food has a distinctly British flavour, with British bread, milk, bacon and sausages being a few of the items imported and advertised for sale.

Lastly, globalization means that Western tourists increasingly feel at home almost anywhere in the Western world because of mobilities of circulating images and global brands of foodstuff, bars, eating places, culture, entertainment, international hotel chains and car rental as well as international systems of credit cards, online booking, money exchange, and so on. Thus, circulating brands and international systems, which are typified by their predictability across time–space, make the world familiar, predictable and repetitive and, in turn, travel-able. As a consequence, there has been an upsurge in individualistic travel patterns, and even package tours are less organized today than they used to be in the heyday of organized mass tourism when globalization, or what George Ritzer calls McDonaldization, was less extensive. The globalization and McDonaldization of the wider society has made the need to McDonaldize the package tour less important. To cite Ritzer and Liska (1997: 102):

> But the more important argument to be made here is that today's tours are less McDonaldized than their procedures precisely because, at least in part, of the very success of McDonaldization. That is, it is because much of the larger society has been McDonaldized that there is less need to McDonaldize the package tour itself. Take standardized meals, for example. In the past, one reason that tour operators had to offer standardized meals was that the food available at any given tourist site was like to prove too unusual and unpredictable and therefore unpalatable for many tourists. However, now tourists can safely be left on their own at most locals since those that want standardized meals will almost undoubtedly find them readily available at a local McDonald's, or at an outlet of some other international chain of fast-food restaurants.

Contra Lash and Urry, Ritzer explains the turn towards more individual and flexible tourism forms as an outcome of 'global McDonaldization' rather than changes in tourists' preferences.

And yet we may also find foreign places familiar because we have become accustomed to, for instance, exotic foodstuff and ethnic restaurants (often imported and run by immigrants). Because of globally circulating 'food mobilities' that displace foodstuffs, national dishes and recipes beyond their national boundaries (S. Gibson, 2007), we can eat the 'foreign', be adventurous cosmopolitans, while at home and 'home bodies' on the move.

Connecting home

The last decade or so has seen a massive proliferation of mobile technologies, most notably digital cameras, laptops, mobile phones, iPods,[1] Game Boys[2] and portable DVD players. They have all made it easier and more common for people on the move in their daily lives to work, communicate, access and surf the internet, play games, hear music and watch movies, and so on. Much research shows that mobiles have become ubiquitous everyday tools, across traditional sociological variables such as class, gender and increasingly also age, more or less always at hand, even, or perhaps especially, when we leave the house (see, for instance, Larsen *et al.,* 2008).

Whereas tourists used to be in more or less incommunicado, the invention and rapid ubiquity of mobile phones, laptops, emails, blogs, social networking sites and access points to the internet mean that connected tourists can retain their virtual proximity with their absent ties at home, out of choice, habit or obligation to others, through what Urry (2002a) terms 'communicative travel'.

An Australian study by White and White (2005, 2007) among foreign campers travelling for substantial periods found that their mobile phones travelled with them and were occasionally put to use as essential connecting devices to family and friends back home. The campers switched off their mobiles most of the time and seldom made calls, to avoid international roaming charges, but cheaper text messages were frequently sent to keep contacts alive (the act of keeping in touch being more important than what is being said). Longer telephone calls were rare and mainly took place when domestic problems had to be talked through or greetings (e.g. birthday) had to be delivered.

Now that airports, camping sites, hotels and many cafés offer wired or wireless internet connection, tourists can easily connect with home and network ties elsewhere on the move though emailing, blogging, home-paging, Facebook-ing and so on. The above-mentioned studies by White and White, and especially Molz's (2004, 2008) virtual ethnography of blogging and emailing among 'round-the-world travellers', show the significance of the internet in conditioning connections to home and making a virtual home (home page away from home!) with a permanent address when one is always on the move. As Molz says:

> Whereas round-the-world travelers move through many destinations during the course of their trip, their Web site address remains a fixed point. Insofar as the Web constitutes a kind of virtual home, travelers can feel at home anywhere as long as they can get online. In contrast to common evocations of the Internet as a metaphor for flux and mobility in a global world, the home page becomes relatively immobile vis-à-vis the travelers' corporeal mobility. Here, we see how home is evoked through the interplay between movement and stasis: the website's *immobility*, a place where the traveler can be located and contacted, becomes a condition of the traveler's *mobile* connectivity.
>
> (Molz, 2008: 7)

Mobile communication technologies are doubled-edged swords that simultaneously allow contact with absent others as well as monitoring by absent others. They allow for a sense of presence at a distance that allows the traveller to be always available, and therefore always under surveillance (Molz, 2006a). On the one hand, the above-mentioned studies indicate that tourists who travel for longer periods appreciate that mobiles and the internet allow them to conduct family life and friendship at a distance in more or less real time. To cite the conclusion of White and White (2007: 100):

> For most of the people interviewed, the establishment and maintenance of copresence was integral to the travel experience. Apart from the small minority of interviewees for whom travel offered a clear break from contact with parents and friends, most of the interviewees made systematic efforts to keep in touch with friends, family members, and colleagues using the various communications services at their disposal.

And yet the studies also show how virtual co-presence at times can be annoying, oppressive and disruptive. Those who desire to use travel to disconnect find the obligation to answer phone calls and emails as disruptive since they can never really leave home and work. For instance, they are reminded of and need to respond to problems at work or home; they may connect because of obligations to friends and family partners, because those left behind feel a need to watch and travel along with them, and this can be felt as a form of surveillance. So the dark side to this connected travel, from the point of view of travellers, is that connectivity in tourism can feel like obligation and surveillance (Molz, 2006a).

Home-making

This chapter has shown how issues of home are central to tourism research once we are concerned with embodied everydayness and dwelling in tourism. Tourists leave behind their usual fixed home, but there is more to home than this physical entity; as tourists we can take our home with us through embodied gestures, routine practices, social habits and small daily rituals; and we must *always* make ourselves at home in foreign places. And, as we have seen, some tourism spaces are designed and regulated in such a fashion that they afford specific forms of home-making: *Oriental* home-making (e.g. enclavic tourism spaces in India and beyond); *international/McDonaldized* home-making (e.g. international hotel chains); and *national* home-making (e.g. nationally themed restaurants and bars). So as Molz's (2008: 2) says: 'The question, then, is not whether home matters anymore amidst all this mobility, but rather *how* home matters'.

A tourism lens attuned to the everyday, communication flows and mobility must also explore tourism performances of 'home-making' (Molz, 2008). By 'home-making' we refer to two distinct embodied and technologized performances. First, there is the embodied performance of making one at home in a foreign place, making a place 'homey'. Second, whereas tourists used to be more or less incommunicado, the invention and rapid ubiquity of the internet and mobile phones mean that connected tourists can retain their virtual proximity with their absent ties at home, out of either choice, habit or obligation to others. Both performances are in part constituted though mundane material objects and we therefore explore what material objects tourists carry with them to create homey sentiments and connections in midst of disconnectivity.

Conclusion

This chapter has documented the need to de-exoticize tourism theory, not to dispense with the exotic and extraordinary as such, but to make space within the theory for 'everydayness'. It was first argued that everyday life should be central to future tourism research because it is a multifaceted notion referring both to routines, ordinary objects and familial faces as well as to excitement, creativity and small-scale disruptive 'tactics'. By incorporating an everyday life perspective into tourism theory it is possible to produce complex, dynamic and contextual accounts of tourism.

In particular, this chapter has outlined how everyday routines and habitual dispositions influence tourism performances that nonetheless still have potentials for creativity and the unexpected. Moreover, we have shown how many tourism performances revolve around pleasant sociality with co-travelling significant others.

Although this chapter has explored connections between everyday life, significant others and travel, there is still a great need for research in this field, and the subsequent chapters of this book explore some of these. We pay attention to the fabric of everyday practices of real holiday experiences to obtain a better idea of what tourists do when holidaying, and how it both ties into *and* occasionally departs from the lived everyday life at home. This includes understanding how tourists might 'discover' as much about their own culture as the one they tour. The everyday characteristic of the tourist spaces passed through and dwelt within also requires more attention. In Chapters 5–8 we show how tourist and everyday spaces are enacted in a blended geography produced and mediated by virtual and material circulations between home and away. There has been an obsession with places that are extraordinary, exotic and inscribed through signs as tourist places. Future ethnographies, as we discuss in detail in the next chapter, need also to take place in ordinary tourist places, and this includes places typified more by 'global flows' than by the 'local' culture, such as McDonald's, Western-style supermarkets and restaurants that *brand* themselves as being Danish, Swedish, Norwegian, Dutch, English, and so on. And we need to follow the flows of emails, text messages, postcards, photographs and souvenirs that tourists make, produce, purchase and circulate to their social networks at home or elsewhere, both while on the move and when at home again. Such mobile ethnographies make it possible to explore how tourist images and objects (re)produce social networks and decorate home geographies.

3 Following flows

Introduction

Chapter 1 described how a performance turn is spreading into and transforming tourist studies. In Chapter 2 we further developed this in a discussion on how performances of tourism and the everyday increasingly blend as travel becomes omnipresent and the spaces of the globe are traversed by travelling bodies and objects as well as virtual, imaginative and communicative networks. We finally observed the need for rethinking studies of tourist performances within a broader transdisciplinary field of 'mobilities studies'.

Although discussions of performance have sparked rich conceptual discussions, they have made fewer methodological innovations. The aim of this chapter is to consolidate this performance turn in tourism studies, and the 'mobilities turn' in the social sciences more generally, by making a *methodological* contribution. The chapter serves two aims. The first is a general discussion of how it is possible to conduct empirical studies of mobile network societies and mobile tourism performances more specifically. The second aim is to outline how this book's mobile tourism ethnographies, in and between Denmark, Turkey and Egypt, are designed and performed.

We argue that 'ethnography' is well suited for studying tourism performances because this method enables sustained observations and accounts of how they take place corporeally, materially and socially within their specific contexts. But we also make the case that there is a need to 'mobilize' ethnography and extend the scope of ethnography to include 'travelling objects' and 'connecting' communications.

The justification for doing this is that existing 'ethnographies' have proved successful in analysing *human* performances within particular sights/sites, but have often been unsuccessful in following flows of people, signs and objects between and across specific places. Most ethnographies are *single*-site ones and overly concerned with humans. Ethnographies of tourism performances have largely neglected the networked mobilities of objects, images, texts and technologies that permit tourism performances to take place, be represented and (re) circulate across often great distances at various sites and times. They have also been somewhat 'a-mobile', as they have mainly observed passing flows within *single* sites. This chapter 'mobilizes' tourism ethnographies by promoting and bringing together what Marcus (1995) calls 'multisited ethnographies' and Urry (2007) more recently 'mobile methods'. These two methods have similarities, although Urry does not discuss 'multisited ethnography' as a source of inspiration for 'mobile methods'. Both are concerned with 'following flows' of diverse mobilities, of people, objects, images, place myths, and so on, in and across multiple sites, in order to highlight how local performances and places are in part constituted through distant flows and mobilities. Both have an agenda of mobilizing the social sciences in order to overcome sedentary approaches to places and dwelling without at the same time promoting a nomadic metaphysic. Inspired by this we develop what we term 'mobile tourism ethnographies' and discuss what implications mobile tourism ethnographies have for the study of performances of tourists and the objects and images that travel along with them.

Ethnography

The performance turn destabilizes semiotic readings in which places and objects are seen as signifying social constructs that can be unveiled through authoritative cultural readings rather than in terms of how they are used and lived with in practice. Ethnography is appropriate for conducting studies of tourist performances because one of the central commitments of this method is 'to be in the presence of the people one is studying, not just the texts or objects they produce' (Miller, 1997: 72). Observations of events as they unfold is characteristic of ethnography. It is a method that requires co-presence with the 'performing people' under study, hence the term *participant* observation. Ethnography allows 'naturalistic' studies of how tourism performances take place as they occur in their natural settings. Much qualitative research in contemporary sociology and geography relies more or less solely on interviews (M. Crang, 2002).

Even in studies of embodied practices, 'what really matters is *talk* . . . talk is made to stand in for all the complexities and subtleties of embodied practice' (Latham, 2003: 1999). Ethnographers and non-representational researchers do not trust interviews on their own: they also examine what and how humans *do* things, corporeally, socially and in conjunction with non-humans (Herbert, 2000). As Erving Goffman says: 'I don't give hardly any weight to what people say, but I try to triangulate what they're saying with events' (1989: 131). Or, to cite Daniel Miller: ethnography 'evaluate[s] people in terms of what they actually do, i.e. as material agents working with a material world, and not merely what they say that they do' (1997: 16–17). This is partly because there can be significant differences between what people 'do' in practice and what they say they do in interviews and partly because most everyday practices take the form of habit, derived in practice. Much social life is conducted unintentionally and habitually. Humans seldom 'think-to-act' (Thrift, 1999: 297). 'People know how to behave "on the move" even if actually how the behaving is to be performed is hard to articulate' (Urry, 2007: 38). This also explains why many interviewees' accounts of their everyday practices are ambiguous, incomplete and sometimes almost lifeless (Latham, 2003). Compared with qualitative interviews, observations better capture the bodily, enacted, technologized and 'here-and-now' quality of practices because they focus on immediate physical doings and interactions rather than retrospective and reflexive talk about how and why such performances took place, and what they mean. Yet interviews that allow space for the unexpected and people's accounts of how their performances are meaningful are vital to avoid portraying 'performers' as 'cultural dopes' (as has been the case in much writing on tourism). Although this book argues for ethnographic observations, it is crucial that this method is a *supplement* to qualitative interviews.

Traditional ethnographies tend to take place within a clearly 'localized' site or field. Ethnographies both by sociologists of urban and community life and by anthropologist of distant cultures have privileged roots over routes, dwelling over travelling. As Clifford (1997: 67) says with regard to the latter: 'In the disciplinary idealization of the "field", spatial practices of moving to and from, in and out, passing through have tended to be subsumed by those of dwelling'. Once arrived, anthropologists became 'home bodies abroad', who privileged face-to-face inter- actions and neglected the traffic and communication connecting the site to the outside world. This is why traditional ethnography can be said to be 'a-mobile' despite the journey to the site. Wittel sums up:

> Long term participant observation in a locally limited area privileges face- to-face relationships and tends to overlook forms of interaction that are more mediated. It privileges permanent residence and tends to overlook movement. It privileges boundaries and thus difference and tends to overlook connec- tions and connectivity.
>
> (Wittel, 2000: unpaginated)

And, more broadly, successive studies of families, communities and social capital 'have followed this steer in taking *close* to mean *near* or interacting frequently face-to-face; and, by extension, significant, important, meaningful' (Fennell, 1997: 90). Social science thus tends to focus upon ongoing and direct social interactions between peoples and social groups that constitute a proximate social structure. Much social science research ignores the movement of people for work, friendship and family, leisure and pleasure. Despite the fact that 'natives, people confined to and by the places to which they belong, unsullied by contact with a larger world, have probably never existed' (Appadurai, 1988: 39), the social sciences mostly fail to examine how social life presupposes both the actual and the imagined movement of people, objects, images and so on (Larsen *et al.*, 2006; Urry, 2007).

That traditional ethnography and other qualitative methods tend to overlook movement, connections and connectivity is precisely why we need to rethink, not dispense with, ethnography, to make it, so to say, more 'mobile'. In specific rela- tion to tourism, the limitation of the performance turn and its use of ethnographic- inspired methods is the restriction of its analytical gaze to particular places (most often famous attractions) and the performances that unfold within them (see tourism ethnographies by Edensor, 1998; Bærenholdt and Haldrup, 2004; Larsen

2005; Obrador-Pons, 2007). While recognizing that localized tourism performances are framed by and draw upon global flows (of stories, objects, people, images, materials and so on), most ethnographies of tourism performances have not yet departed from the deep-rooted anthropological idea that 'ethnographies' take place *within* bounded sites. Being based on '*single*-site ethnographies', these studies do not 'follow flows' between sites. In fairness it should be noted that some studies emphasize how tourist places are made and remade as they are 'toured' by particular modes of mobility, cultural scripts and embodied performances. Indeed, Mordue (2001: 184) notes that 'the major benefit of [this] ethnographic research is that it allowed an in-depth investigation of how local tourism performances are specific, yet mediated by global processes'.

Yet it is still the case that most tourism ethnographies are incapable of capturing the role of heterogeneous flows in enabling particular sites. Below we suggest how so-called 'multisited ethnography' and 'mobile methods' might remedy this. We start by discussing how much social theory at the end of the twentieth century emphasized the increased mobility of objects, people, information and meanings, and how this transformed the world into a 'single field of persistent interaction and exchange' (Hannerz, 1996: 19) and the need to develop methodological frameworks for dealing with 'the global', 'the mobile' and 'network societies' as emerging new social realities.

A global mobile world

One of the first sociologists to deal with 'the global' as an emerging new social reality was Anthony Giddens. Giddens (1990: 64) defines globalization as the intensification of worldwide social relations, which link distant localities in such a way that local happenings are shaped by events occurring many miles away and vice versa. Drawing on his structuration theory (Giddens, 1984), he argues that individual places are connected to and always haunted by other places, even if this is not necessarily visible on the scene:

> Locales are thoroughly penetrated by and shaped in terms of social influence quite distant from them. What structures the locale is not simply that which is present on the scene; the 'visible form' of the locale conceals the *distanciated relations*, which determine its nature.
>
> (Giddens, 1990: 19)

The nature of a particular place is thus always in part determined by 'distanciated relations' with other places: a 'trivial act may trigger events far removed from it in time and space' (Giddens, 1984: 11). Such dualistic thinking of the global and the local highlights the limitation of studying places through traditional single-site ethnography, by erecting walls and neglecting connections and movement in and out of the site.

Another highly influential account of the modern society's global and mobile nature is that of John Urry (see also Chapters 1 and 2). In major books such as

Sociology beyond Societies (1990) and the recent *Mobilities* (2007), Urry argues that twenty-first century sociology ought to move beyond societies and engage with the various mobilities – of people, objects, signs, risks and so on – that move in and across societies and, in doing so, connect and disconnect them from each other. Urry makes the case that sociology has overly focused upon ongoing geo- graphically propinquitous communities based on more or less face-to-face social interactions with those who are co-present. But now, travel, flows and connections rather than bounded societies ought to be the subject matter of sociology, because 'it sometimes seems as if all the world is on the move' (Urry, 2007: 3). The world is on the move because there is an immense increase in the scale of (ibid.: 47–8):

1   *physical travel* of people for work, leisure, family life, and migration, to meet people face to face, to see a place directly or experience/attend social event that can not be missed;
2   *physical movement* of objects (consumer goods, souvenirs, etc.);
3   *imaginative travel* elsewhere through images, television and memories (arm- chair travel);
4   *virtual travel* on the internet (cybertourism) and digital movement of images, messages and information;
5   *communicative travel* via text messages, telephones, emails, Skyping, etc.

Most social research focuses upon one of these separate mobilities, such as pas- senger transport or mobile telephony or the internet, and generalizes from that. Urry, in contrast, examines the interconnections between these different mobili- ties, of humans and non-humans, in material or digital form. What distinguishes Urry's work from mainstream transport and tourism research is the idea that studies of physical movement of people and objects must be supplemented with studies of imaginative, virtual and communicative travel, 'how the transporting of people and the communicating of messages, information and images may overlap, coincide and converge through digitised flows' (Urry, 2007: 8). Urry's suggested mobilities paradigm

> emphasizes how all social entities, from a single household to large scale corporations, presuppose many different forms of actual and potential move- ment. The mobility turn connects the analysis of different forms of travel, transport and communications with the multiple ways in which economic and social life is performed and organized through time and across various spaces.
>
> (ibid.: 6)

Not unlike Giddens, Urry notes how movement and connections are crucial in making places. Thus, Urry speaks of 'places of movement'. Following Hetherington (1997), he argues that places are not like islands – they are not rooted in one place and they do not exist autonomously. They are rather like ships. Places are travelling, constructed through, as Clifford (1997) would say, routes as well as roots. Urry does not see places as unique, bounded and fixed islands; rather they come into existence through relationships. Places float around within mobile, transnational networks of humans, technologies, objects and images that continuously connect and disconnect them to other places (Urry, 2007: 42). Or, as Dorren Massey (1994: 217) puts it: 'what gives a place its specificity is not some long internalized history but the fact that it is constructed out of a particular constellation of relations articulated together at a particular locus'. Communities are impure and porous. Travel is central to communities, even those characterized by relatively high levels of apparent propinquity and communion.

The mobilities paradigm is 'a movement-driven sociology' exploring connections between various mobilities, and how actual, potential and blocked mobility is constitutive of distanciated and networked 'social life', from family life to large organizations (Urry, 2007: 43). The mobilities paradigm is also a 'post-human turn'. In the spirit of actor network theory, it represents a critique of humanistic accounts that separate humans and non-humans, social worlds and material worlds, and therefore understand societies as purely social entities (ibid.: 45). If the social world consisted solely of 'naked bodies' and pure face-to-face interaction, all the world would *not* seem 'on the move' for, 'left to their own devices, human actions and words do not spread far at all' (Law, 1994: 24). Duim highlights this fact in relation to what he calls 'tourismscapes':

> An array of networked objects, media, machines and technologies extend tourismscapes in time–space . . . take away the planes, travel books and brochures, maps, timetables, the internet, passports, or internationally accepted ways of payments, and time–space *de*compresses immediately.
>
> (Duim, 2007a: 968, our italics)

Modernity's defining 'time–space compression' (Harvey, 1989) and 'time-space distanciation' (Giddens, 1990) thus depend upon a vast array of actor networks in which non-humans also act: machines, technologies, objects, signs, information and so on. The mobilities paradigm presupposes that the powers of humans are always enhanced by various non-humans, such as technologies and objects, and mobility systems.

Mobile communications also mean that much everyday life is performed face to interface rather than purely face to face. New communications afford new forms of social interaction at a distance:

> The development of new media and communications does not consist simply in the establishment of new networks for the transmission of information between individuals whose basic social relationship remains intact. Rather, the development of media and communications creates *new* forms of action and interaction and new kinds of social relationships – forms that are different from the kind of face-to-face interaction which has prevailed for most of human history.
>
> (Thompson, 1995: 81)

Larsen *et al.* (2006) discuss how the idea of an uninterrupted face-to-face sociality, disentangled from technological devices, is becoming rare. Face-to-face meetings transform into face-to-interface interactions when computer documents are worked upon, PowerPoint presentations begin, mobile phones ring, text messages arrive, and so on. Face-to-face meetings are mediated and always connected to other meetings; they are typified by 'absent presence' (Gergen, 2002). As Callon and Law (2004: 6, 9) maintain more generally, 'presence is not reducible to co-presence . . . co-presence is both a location and a relation'.

> To inhabit such machines is to be connected to, or to be at home with, 'sites' across the world – while simultaneously such sites can monitor, observe, and trace each inhabited machine . . . others being uncannily present *and* absent, here and there, near and distant, home and away, proximate and distant.
>
> (Urry, 2004: 35)

This ties into Manual Castells's hugely influential idea of the 'network society' (Castells, 1996, 2000, 2004). This is a society made up of networks that are powered by microelectronics-based information and communication technologies:

> What is specific to our world is the extension and augmentation of the body and mind of human subjects in networks of interaction powered by micro-electronics-based, software-operated, communication technologies. These technologies are increasingly diffused throughout the entire realm of human activity by growing miniaturization [and portability].
>
> (Castells, 2004: 7)

In particular, 'networked computers', 'mobile telephony' and fast transportation have been crucial in producing this global network society of 'timeless time' and 'spaces of flows'. The latter highlights how dealing with distances and networking at a distance have become crucial:

> Simply put, the *space of flows* is the material organization of simultaneous social interaction at a distance by networking communication, with the technical support of telecommunications, interactive communication systems, and fast transportations.
>
> (Castells *et al.*, 2007: 171)

Castells's account is a macro one, and he does not address the network society in relation to friendship and family life or the everyday (but see Castells *et al.*, 2007). Yet this notion suggests that 'sociality' and everyday life are increasingly networked with mobile communications, and performed through interfaces and phonescapes rather than purely face to face. Western households and individuals are increasingly networked. People are plugged into an ever-expanding array of communications that connect them to one another and to the outside world. These include postal systems, radio, television, satellite television, landline phones and, more recently, mobile phones, computers, digital cameras, email accounts, home pages, blogs and social networking sites. For instance, 85 per cent of Danish households have one or more computers, and 80 per cent are connected to the internet.[1] Thirty-one per cent of the Danish population (as of December 2008) have a personal Facebook profile.[2] The home has become a communication and information hub (Wellman *et al.*, 2006).

And 'individuals' are themselves 'networked', which is particularly evident with regard to mobile phones, which afford 'person-to-person' connectivity, or 'networked individualism' that 'suits and reinforces mobile lifestyles and physically dispersed relationships' (Wellman, 2001: 239). 'The person has become the portal' (ibid.: 238). Widespread mobile phone ownership enables connectivity in the midst of absence, distance and disconnection (Licoppe, 2004; Larsen *et al.*, 2006). Phone calls on the move and impromptu text messages have become crucial networking practices in many countries around the world, especially in Europe. In 2004, 9 out of 10 people in the EU countries (and Norway) were mobile phone users (Castells *et al.*, 2007). And many individuals are networked with virtual sites; they network virtually and their 'profile' is available online. The last couple of years has also seen an astonishing increase in networking on the internet – measured in terms of users of social networking sites (see Wikipedia's lists of numbers of users of 'notable social networking sites').[3] As their names reveal, such sites are primarily about networking with old and new friends, and they are primarily – but not exclusively – for teenagers and youngish people.

The significance of such mediated and mobile networking means that *access* to communication technologies, transport, meeting places *and* the social and technical *skills* of networking is crucial for sustaining social networks and hence general social wellbeing. This is what Urry calls 'network capital'. In societies organized around 'circulation' and 'distance', the greater the significance of network capital within the range of capitals available within a society (Urry, 2007: 52):

> Network capital is the capacity to engender and sustain social relations with those people who are not necessarily proximate and which generates emotional, financial and practical benefit . . . Those social groups high in network capital enjoy significant advantages in making and remaking their social connections, the emotional, financial and practical benefits.
>
> (ibid.: 197)

Urry is interested not in mobility technologies in and of themselves but in what they *afford* to social networks, how they *potentially* change them:

> What are key are the social consequences of such mobilities, namely, to be able to engender and sustain social relations with those people (and to visit specific places) who are mostly not physically proximate, that is, to form and sustain networks. So network capital points to the real and potential social relations that mobilities afford. This formulation is somewhat akin to that of Marx in *Capital* where he focuses upon the *social* relations of capitalist

production and not upon the *forces* of production *per se* (1976). My analogous argument is that it is necessary to examine the social relations that the means of mobility afford and not only the changing form taken by the forces of mobility.

(ibid: 196)

Network capital therefore comprises (ibid.: 197–8):

1　an array of appropriate documents, visas, money, qualifications;
2　others (workmates, friends and family members) at a distance;
3　movement (and communication) competences;
4　location-free information and contact points;
5　communication devices;
6　appropriate, safe and secure meeting places;
7　physical access to cars, road space, fuel, lifts, aircrafts, trains, ships, taxis, buses, trams, minibuses, email accounts, internet, telephones, and so on;
8　time and other resources to manage and coordinate nos. 1–7.

Thus, network capital complexly comprises technical, cognitive and social skills and depends on 'accesses' to various documents, strong and weak ties, communication technologies, databases *and* transport technologies. Whereas some of these 'accesses' depend upon appropriate 'economic capital', others are not necessarily economic in nature. For instance, many 'poorer' migrants are likely to have many 'others' at a distance (no. 2). Likewise, whereas 'access' to communications and travel requires 'economic capital', what Urry terms 'competences' (no. 3) is largely independent of this type of capital. So there is no linear or direct proportionality between 'economic capital' and 'network capital' (Larsen and Jacobsen, 2009). The notion of network capital resembles Kaufmann *et al.*'s (2004) notion of 'motility' that refers to *potentials* for actual mobility. Motility 'encompasses interdependent elements relating to access to different forms and degrees of mobility, competence to recognize and make use of access, and appropriation of a particular choice, including the option of non-action' (ibid.: 750). As people are distributed 'far and wide', so network capital is essential for social life: 'Without sufficient network capital people will suffer social exclusion since many social networks are more far-flung' (Urry, 2007: 179). Whereas Bauman's notion of 'liquid modernity' (Bauman, 2000) suggests the end of immobile hardware, mobile network societies are in fact full stationary platforms. This is particularly clear in Urry's writing. Urry works with a dialectic of mobility and immobility as flows of mobility presuppose 'immobile' material worlds, multiple fixities or moorings, especially platforms, such as coaxial cable systems, satellites for radio and television, mobile phone masts, roads, garages, stations, aerials, airports, docks and internet cafés. 'There is no linear increase in fluidity without extensive systems of immobility' (Urry, 2007: 54).

Following the work of Giddens, Castells and especially Urry, we have described contemporary Western societies as mobile networked societies. Now we discuss how we can research this type of 'society' empirically, how we can 'follow flows', of people, things, images and so on, in, through and across 'moorings', beginning with discussions of multisited ethnography within anthropology.

Multisited ethnographies

As James Clifford (1997) famously argued, ethnography needs to leave behind its preoccupation with discovering the 'roots' of cultural and social forms and instead trace the 'routes' that produce and reproduce them. Following Arjun Appadurai, George Marcus has argued that the investigation of increasingly interdependent and fluid phenomena makes it necessary to 'move out from the single sites and local situations of conventional ethnographic research designs to examine the circulation of cultural meanings, objects and identities in diffuse time–space' (Marcus, 1995: 96). To some extent, multisited ethnography is 'old news' in some parts of the social sciences. For instance, Hannerz (2003) argues that migration researchers have always preferred ethnographic studies at 'both ends' of the migration flow. However, Marcus's more programmatic suggestion of a multisited ethnography of the world system has been important for the use of multisited studies in wider cultural studies of society. According to Marcus, such a multisited ethnography must be 'designed around chains, paths, threads, conjunctions, or juxtapositions of locations' to follow people, things, metaphors, stories, lives and conflicts in motion. And he continues:

> In short, within a multi-sited research imaginary, tracing and describing the connections and relationships among sites previously thought incommensurate is ethnography's way of making arguments and providing its own contexts of significance.
>
> (Marcus, 1998: 14)

Unlike traditional ethnography, which defines sites as *material* dwelling places, multisited ethnography also deals with *virtual* sites such as databases and blogs, not in an isolated cyberspace, but in relation to physical everyday places such as internet cafés, work places and private living spaces, as virtual worlds and material worlds are not separate entities (Wittel, 2000). Multisited ethnography privileges routes rather than roots: connections and networks. Following Wittel, it represents a move from ethnography of fields (a geographically defined locality) to 'ethnography of networks':

> Networks are still strongly related to geographical space – like field. Unlike field, a network is an open structure, able to expand almost without limits and highly dynamic. And even more important: A network does not merely consist of a set of nodes, but also of a set of connections between the nodes. As such, networks contain as much movement and flow as they contain residence and localities. An ethnography of networks would contain the examination of the

nodes of a net and the examination of the connections and flows (money, objects, people, ideas etc.) between these nodes.

(Wittel, 2000: unpaginated)

Multisited ethnography is less about sustained dwelling in one field than it is about following the flows of humans and non-humans in and across a particular field or several fields, so the ethnographer needs to travel a great deal, physically and through communications: it is mobile ethnography par excellence.

This focus upon networks, connections and flows also means that multisited ethnography and the methodology of actor network theory are related. Actor network theory is a descriptive method that follows the relational practices of actor networks. It involves tracing the 'footsteps' of the hybrids under study and describing their ways of accomplishing the field by weaving together various places and actor networks (see Chapter 4). Actor network theory is clearly multi-sited by arguing that 'relations in and out of the field in question are as important as what goes on within it as these constitute the field to much extent and hence, the boundaries of a field is primarily a practical achievement' (Johanneson and Bærenholdt, forthcoming: unpaginated; Latour, 2005).

Multisited methods may here provide helpful tools for uncovering tourism's many 'sites of production', investigating the various settings in which tourist materials and meanings are consumed and produced, and highlighting the complex combinations between sites, meanings and materials. A multisited approach can help us think of tourism as taking place not between spaces of home and spaces of leisure but in networks. This approach resembles what Joy Hendry (2003: 510) calls 'globology', i.e. studies 'identifying and describing discourses held by people with different ways of defining themselves but who communicate through new global forms of technology and exist only because of these forms of technology'. Alongside migration and electronic media, tourism is one of the most significant forces in transforming the globe into one coherent field of inter-action. And, like migrant cultures, tourism can be seen as a culture of circulation and connections. Both

transgress the boundaries of home and away, well-known and imagined, by creating specific irregularities because both viewers and images are in simultaneous in circulation. Neither images nor viewers fit into circuits of audiences that are easily bound within local, national or regional spaces.

(Appadurai, 1999: 4)

And, like migrants, tourists are also parts of networks and circuits that are not easily located within national, local or regional spaces but encompass both local-ized performances in place as well as global processes. Moreover, such circuits also break down the barrier between the known and the fantasized. As Appadurai (1999: 7) puts it: 'imagination is today a staging ground for action, and not only for escape'. We now discuss how multisited ethnography has been used by schol-ars within consumption studies.

Consumption studies and the global culture industry

In consumption studies, multisited approaches have produced interesting studies of the 'traffic in/of things' in relation to transnational commodity chains (Jackson, 1999; P. Crang *et al.*, 2003). For instance, Cook *et al.* (2004) trace the flow of ordinary consumer goods such as papaya fruit and West Indian hot pepper sauce through complex transnational networks between consumers, producers and retailers. Other multisited studies trace the production and consumption of green beans, tourism souvenirs and 'global news' (Friedberg, 2001; Ateljevic and Doorne, 2003; Hannerz, 2003; Cook and Harrison, 2007).

Such multisited studies of things-in-motion uncover some of the different meanings and effects that objects can have in different places and on the move. They de-fetishize apparently trivial consumer goods by showing the 'material links that cut across boundaries between people' (Hendry, 2003: 499) and highlight the importance of recognizing the distinct and variant effects and attitudes related to such movements in different social and cultural contexts. In doing so, these studies provide a more reflexive research praxis by shifting attention from the inherent meanings of objects, places, images and texts to the contested 'productions of various representations as moments for situated reading and interpretation by all actors (including the researcher!)' (M. Crang, 2005: 227).

Lash and Lury's recent *Global Culture Industry* (2007) is another stimulating example of multisited research concerned with the travel of *objects* through diffuse times and spaces (see also Chapter 1). Over a period of time, this book follows the flows of seven cultural objects (for instance brands such as Nike and Swatch as well as movies such as *Trainspotting* and *Toy Story*) as they travel in and through great many sites, as both signs and objects. Lash and Lury argue that the cultural industry has become a global one of circulation, in which media become things and things become media. They speak of a 'mediation of things' (ibid.: 8–9). Their methodology is one of 'following the object' and the field is the spaces travelled by the objects through their life courses. So this is a mobile ethnography concerned with the social life of objects:

> As the study developed, we too came to think of our objects as having a life. We were using an anti-positivist, a *humanist* method: but what was involved was a humanism of the inhuman. We were involved in a mobile ethnography (in the very broadest sense of the word), in which the ethos was a community of things.
>
> (ibid.: 20)

Mobile methods

Although there is much theoretical discussion of multisited research, there is scant guidance on how such research can be performed in practice. We suggest that Urry's discussion of mobile methods can be helpful in this regard. Surprisingly, Urry (2007) does not mention multisited ethnography as a source of inspiration, even though there are many overlaps between his mobile methods and multisited ethnography. Urry argues that the mobilities paradigm involves new kinds of methods: 'What I establish here is that research methods also need to be "on the move", in effect to simulate in various ways the many and interdependent forms of intermittent movement of people, images, information and objects' (ibid.: 38). Mobile methods include, according to Urry (ibid.: 40–2):

1   Direct or virtual *observations* of humans' movements and their face-to-face interactions with places, events and other people.
2   *Mobile observations* of people or objects as they travel along. This requires that the researcher follows in the footstep of such mobile entities in and across various sites.
3   Time–space *diaries* in which humans record what they are doing and where, how they move during those periods and the modes of movement. Here the inspiration is 'time geography' rather than ethnography.
4   The use of 'virtual ethnography' to explore the imaginative and virtual mobilities of people through analysing text messages, websites, multiuser discussion groups, blogs, emails, listserves, and so on.
5   The travel of objects. This involves studying the cultural biographies or social life of objects, while the geographical flows of messages and people can be tracked through the use of global positioning systems.
6   The conceptualization of places as being on the move rather than fixed. 'Such hybrid systems that contingently produce distinct places need examination through methods that plot, document, monitor and juxtapose places on the go'.

Mobile ethnographies of tourism

In the following we suggest the notion of 'mobile tourism ethnography', which combines insights from mobile methods and multisite ethnography. Mobile tourism ethnographies research how movement and connections (and their associated moorings), of people, objects, images and myths, are crucial in making tourism performances and places: 'destinations of tourism are "hollowed out" and not self-present' (M. Crang, 2006: 58). This is in contrast to traditional tourism theory, in which there is a

> presumption of not only a unity of place and culture, but also of the immobility of both in relation to a fixed cartographically coordinated space, with the tourist as one of those wandering figures whose travels, paradoxically, fix places and cultures in this ordered space.
>
> (Lury, 1997: 75)

Mobile tourism ethnographies see performances as constructed through routes *and* moorings, connecting home and away as well as physical, object, imaginative, virtual and communicative mobilities. A mobile tourism ethnographies

approach to tourism performances would then involve tracing the multiple mobilities and moorings that make tourism happen across often geographical dispersed
spaces. The agenda of mobile tourism ethnographies is as follows.

First, analyse the production and circulation of place myths though images,
objects and rumours, and how tourism places are affected by faraway place
myths. Following Giddens, we may say that these imaginations are not necessarily
present in place but they have *effects* in place. As Mike Crang (2006: 49) argues:

> tourism works as interplay of movement and fixity, absence and presence.
> That is, the tourist seeks to be present at a place, but as we examine those
> places we find that they are shot through by absences where distant others,
> removed in space and time, haunt the sites.

This is particularly evident with cinematic tourism, in which imaginary geographies seen and circulated on film and television choreograph tourists' actual
routes, performances and interpretations in and of places (Mordue, 2001; M.
Crang, 2003; Tzanelli, 2007). This is also the case with political events displayed
and circulated globally on the internet and television, an example being the
publication of the cartoons of the prophet Muhammad in a Danish newspaper
in 2005 and subsequently throughout the world (see Chapters 1 and 5). And yet
it is important that we understand 'that images do not just obscure a true place,
but rather constitute the very sense of places themselves. We have then to look at
the performativity of images and texts moving and making through processes of
signification' (M. Crang, 2006: 54–5).

Second, follow the 'footsteps of tourists', either by asking tourists to 'track'
their movement through time–space in time–geography-inspired diaries or by
physically travelling *along* with them, observing and interviewing them. Both
methods (especially if they are combined) pay attention to how and where tourists
move, what places they visit and perform, both on and off the beaten track. This
might highlight how tourism performances not only take place at famous sights
and involve exotic practices, but also happen across multiple sites (moorings),
ordinary as well as extraordinary, where many mundane everyday practices are
also performed. Traditional single-site ethnographies of tourism performances
at attractions suffer from the fact that most tourists spend very little time there
before moving on. Researchers often end up glancing at passing flows and doing
hurried interviews. As the anthropologist Edward Bruner reflects:

> A key difficulty in studying tourists is methodological – the tourists move so
> fast through the sites that it is hard to keep up with them . . . It is relatively
> easy to begin a discussion but in the middle of a sentence the tour leader
> announces that the group is moving on to the next site, and your inform-
> ant has disappeared . . . I felt that the only way for me to enter into tourist
> discourse would be to join the tour group. As a guide, I would be an insider
> and I could observe how the tourists actually experienced the sites and events
> to which they were exposed.
>
> (Bruner, 2005: 228)

Whereas one way to follow the footsteps of tourists over a sustained period of time is therefore to become a guide, a second is to participate as a tourist on a guided package tour – the strategy of this research project (see next section) as well as H. Andrews's ethnography of Britishness in a Spanish charter tourist:

> However, there is also a degree of role playing, which in this situation means enacting the part of a British tourist – lying on the beach to sunbathe, swimming in pools and the sea, playing bingo, eating fry-ups, and generally joining in tourist activities, such as island tours, market visits, various forms of night-time entertainment including night-clubs, dancing and bar crawls. In short, my relationship with the tourists was as one of them.
>
> (H. Andrews, 2005: 250)

Such a tour permits the researcher close proximity to a group of tourists for a fortnight or so, allowing time for casual conversation on the move, in-depth interviews while resting and sustained observations of the diverse performances and sites that guides and tourists enact, visit and pass through. This means that qualitative interviews can be approached ethnographically, as they can take place while performances are being enacted and involve souvenirs, guidebooks, maps and available communications technologies. Another well-established method of following flows and performances is time–geography. Haldrup (2004) and Latham (2003) both use time–geography-inspired diaries and diagrams to track the spatio–temporal rhythms and styles of city-dwellers and tourists' mobility practices (see Gren, 2001, for a more general, critical discussion of time–geography's contribution to cultural geography).

Third, follow the objects that tourists transport to – and home – from the destination. There is a need to unpack the flows of food, drinks, mobile phones, cameras, toys, literature, music, medicine and clothes that couples and families take along in order to make themselves at home – in the double sense of holding on to one's known world and fitting into a new home, making it 'dwell-able' and safe. The common idea that tourism is the opposite of everyday life also neglects the fact that tourists need to make themselves at home in a strange, foreign place and that they transport some of their home with them. This analysis is particularly significant in relation to Oriental destinations such as Turkey and Egypt because, for many tourists, these places are very different from their home geographies, as well as from traditional mass tourism destinations such as Spain and Greece, and they are likely to be perceived as containing some risks in relation to food, hygiene, medical care and cultural norms for clothing, and so on. What objects tourists take home with them, the photographs and various souvenirs they produce, buy, display and circulate to their friends, also need to be analysed (see below).

Fourth, follow 'communication flows', flows of digital/virtual travel. As discussed, recent years have seen a proliferation in communication technologies that allow tourists to be in more or less constant touch with their absent ties though emails, text messages, photo messages and free voice-over telephony, technologies designed for mobility and geographically dispersed social networks. As discussed above, such communications can blur distinctions between presence and absence, near and far, home and away. Examining what communication technologies travel with tourists and how they are used in practice illuminates tourists' 'connected presence' with people and places at home or elsewhere. Although our mobiles always seem at hand when we conduct our everyday lives, the question remains 'Is this also the case on holidays?' Are tourism experiences characterized by connectivity and 'communicative travel' despite the high price of using them abroad? Or do tourists desire a holiday away from the mobile phone and the culture of connectivity it entails?

Mobile tourism ethnographies need to follow the networked flows of emails, text messages, telephone calls, postcards, photographs, souvenirs and so on that tourists make, produce, purchase and circulate to their social networks at home or elsewhere, both while on the move and when at home again. Following such flows makes it possible to explore both how 'local' tourism geographies circulate in distant places as place myths, and how tourists use travel tales, images and consumer goods to (re)produce social networks and decorate their homes and bodies. Such an approach casts light both on how imaginative geographies are produced and circulated in and through lay geographies and personal representations that crisscross sites between home and away and on how tourist performances are increasingly being staged more or less live for an absent yet co-present audience. For instance, mobile tourism ethnographies should examine how tourist photography today might no longer be directed only at a future audience but instead to a more or less instantaneous live audience, now that camera phones, internet cafés, emails and travel blogs 'timelessly' transport images (technologies permitting!) over great distances. Is the new spatial–temporal order of tourist photography and by implication many other tourism performances, one of 'I am here' rather than 'I was here' (Bell and Lyall, 2005)?

Fifth, ethnographic research needs to be situated not only at distinctive tourist sights, but also in more or less ordinary tourist places (restaurants, swimming pools, supermarkets, and so on). Much tourism theory has been obsessed with places and practices that are extraordinary, exotic and clearly inscribed through signs as tourist places (MacCannell, 1976/1999; Urry 1990; see also critique thereof in Chapter 2). This is the major reason why tourism ethnographies have so far been mainly of attractions. But tourists spend much time outside attractions and engage in various mundane, more or less pleasurable practices, such as eating, socializing, relaxing, shaving, bathing, waiting, shopping, riding on buses and trains, and so on (see Chapter 2).

Sixth, in addition to undertaking ethnographies of destination sites that afford connections to home, mobile tourism ethnographies also need to take place in tourists' private homes, where flows of imaginative geographies – marketing material, political news, souvenirs, photographs and travel tales – are consumed, seen, worn, displayed, performed and disseminated to other households. Pre-travel interviews can reveal some of the Oriental place myths (risks of terror, cultural stereotypes and so on) that travel along with tourists to actual destinations. As discussed in Chapter 2, tourists never just travel to places: their mindsets travel *with* them. We may note how the imaginative geographies of tourism are as much about home as about faraway places. Post-travel research is particularly interesting from a multisited ethnography perspective as it can document how tourism performances affect, and sometimes take place in, sites that are spatio-temporally remote from particular tourist places. At home people perform tourist memories, travel tales and popular imaginative geographies by chatting over souvenirs and holiday snaps displayed on fridges, work desks, mobile phones, computers, and so on. Post-travel home ethnographies need to explore the afterlife – storing, displaying and circulation – of souvenirs, photographs and email postcards, which ties in to the increasing significance that cultural geographers are attributing to home geographies (see Blunt, 2004), including personal photography (G. Rose, 2003). Do souvenirs travel well, or do they change meaning with their movement and displacement? Are holiday photographs more widely exhibited and distributed now that emails transport them timelessly and blogs and social networking sites exhibit them globally? Post-travel research thus destabilizes the common idea in tourism theory and cultural geography that tourism and everyday life/home belong to different ontological worlds. Moreover, post-travel research allows one to explore to what degree 'actual tourism' changes people's perception of the particular place they visit and their world-view more generally. One way to do this is by exploring the souvenirs and personal photographs tourists bring home and disseminate.

The remainder of this chapter discusses how this book uses some of the methods listed above in the tracing of multiple tourist flows between the Orient (Turkey and Egypt) and northern Europe (Denmark) as enacted and circulated by Danish tourists.

Following flows

This book employs mobile tourism ethnographies to explore how (primarily Danish) tourists perform, and make themselves at home in, the Orient, weaving together 'home' and 'away', corporal, material, virtual, imaginative and communicative mobilities, connections and moorings. It traces the flows of images, narratives, people and things-in-motion that script, stage and circulate the Orient *and* touring tourists across multiple moorings and sites. Giving illuminating attention to the fabric of everyday life and mundane, 'homely' practices, the real holiday experience is placed in the foreground as it takes place during and after the journey itself. And yet this book also investigates some of the politics of tourism performances and of the Orient more generally. As discussed in Chapter 1, we connect the everyday and the Orient through the notion of practical Orientalism. We explore ethnographically how guides stage and tourists consume the Orient in the aftermath of September 11, the recent Muhammad cartoon crisis, terror bombings in tourist destinations in Egypt in 2005 and 2006, and anti-Muslim xenophobia in many European countries. What far-reaching flows and historical/national/global place myths are mobilized and demobilized in this process? To what extent do tourists' physical encounters with the Orient and Muslims reproduce or challenge media-circulated place myths? This research thus combines 'travel ethnographies' (at various sites) with 'home ethnographies', which are connected by exploring how internet cafés, mobile phones, social networking sites, blogs, text messages, emails, photographs, telephone calls, souvenirs and consumer goods bridge home and away, absence and presence, being-here and being-there. Mobile tourism ethnographies challenge the conventional idea that tourism research needs to take place at attractions and that home is something that tourists leave entirely at home. Instead they highlight fluid connections between home and away. This also explains why this book is based upon research undertaken both at tourist destinations in Turkey and Egypt as well as in tourists' homes in Denmark. In this book we draw on four main sources of methods and materials collected 'on travel' and 'at home'.

Travel ethnographies

We perform travel ethnography by participating on three package tours organized by Danish travel companies. We participated in two one-week package tours to Alanya, Turkey, and a two-week tour to Sharm el Sheikh, Egypt. Both destinations are highly popular destinations for Danish tourists and many Danes have bought holiday flats in Alanya (Figure 3.1). We conducted research at various sites and followed tourists by participating in guided tours, making observations and striking up casual conversations with our co-travellers at various hotel moorings (such as the morning buffet, swimming pool, bar, internet corners) and attractions, bazaars, markets, and so on. On these tours we participated as fully participating tourists (and we were 'disguised ethnographers' except when we made formal interviews) and our families partook in two of them. Additionally, we carried out a two-week ethnographical study at the major sights of Istanbul, focusing primarily on photography and sight-seeing performances, performing visual ethnography and undertaking a series of shorter and longer semistructured interviews (reported in Chapter 7). Here we stayed in a low-budget hotel catering for independent travellers rather than package tourists. Since we have in part used 'disguised ethnography', we have done our utmost not to report situations or conversations that could disclose the identity of the tourists we travelled and chatted with.

Home ethnographies

We made 18 post-travel home ethnographies (Figure 3.2). These consisted of semistructured interviews lasting between one and two and a half hours and ethnographic investigations of how people use, store and distribute tourist objects and photographs. All interviews have been fully transcribed. Some of the interviewees were people whom we had already interviewed while doing fieldwork, but the majority were recruited through a Danish package tour organizer. We carried out the interviews at people's private homes because we are interested in exploring the afterlife of tourism objects such as souvenirs and photographs.

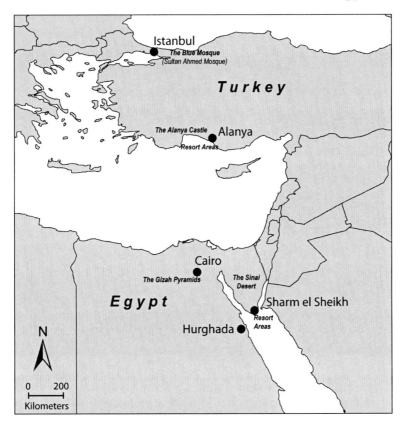

Figure 3.1 Fieldwork sites (Turkey and Egypt).

Prior to the interview we informed the interviewees that we wished to talk about and see their guidebooks, photographs, cameras, mobile phones, postcards and souvenirs, and we asked them to have these as well as their home computer 'at hand' so that we could explore both the content and the organization/display of their holiday photographs. Thus, we tried to construct the interviews around such sets of communicative objects and communications. The interview guide covers various themes such as occupational, educational, marital details; access to communications (e.g. digital camera, blog/social networking site/computer, printer); why they travelled to Egypt/Turkey; how and through what media the trip was organized; their expectations, fears of terror and prior mental images of the Orient; mental geographies of the Muslim world; everyday routines on the destination; spatial patterns; typical activities; interactions with fellow tourists; interactions with the local population; internet use, communication with absent others (emails, text messages, blogging and so on); photography practices; souvenirs; if and how tourism changes attitudes.

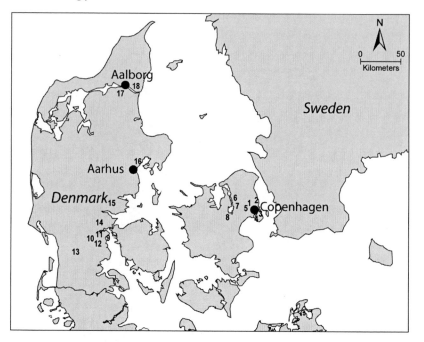

Figure 3.2   Home ethnographies (Denmark). 1, Keith and Lisa (students, 20s), Copenhagen; 2, Ann (post-graduate, 30s) and Pennie, (primary school pupil, < 10), Copenhagen; 3, Joan and Ahmed, education manager/restaurant manager (40s), Copenhagen; 4, Heidi (art student, 20s), Copenhagen; 5, Henrik (curator, 40s), Copenhagen; 6, Leigh and Ann-Lee (retired, 60s), Zealand; 7, Anne and Michael (retired, 60s), Zealand; 8, Jane (retired, 60s), Roskilde; 9, Susan (part-time assistant, 50s), southern Jutland; 10, Jørgen (business manager, 40s), southern Jutland; 11, Lotte, (caretaker, 20s), southern Jutland; 12, Morten and Camilla (manual worker/student, 20s), southern Jutland; 13, Heinz (retired, 60s), southern Jutland; 14, Simon (sales manager, 40s), Karen (translator, 40s), Judith and Lilith (secondary school students, teenagers), southern Jutland; 15, John and Sara (manual workers, 40s), southern Jutland; 16, Susi (kindergarten deputy manager, 30s), Aarhus; 17, Anja (student, 20s), Aalborg; 18, Hanna (self-employed, 30s), northern Jutland.

Diaries

In order to follow the flows of tourists, we also experimented with making and distributing diaries to a small group of tourists before they set off to Egypt or Turkey. The purpose of the diaries is twofold. First, they allow us to follow the corporeal movement of tourists day by day. Second, and most importantly, they invite people to record the activities and experiences that they engage in day by day throughout a week, especially those that they find particularly enjoyable, annoying, strange or hilarious, whether they are supposedly small or large events, mundane or exotic. Each day is divided into time slots and they give the families plenty of space to tell their stories. It is stressed that there is no particular right way to write them and that people are welcome to experiment with the form and include entrance tickets, images, and so on.

Virtual ethnographies

Finally, we studied various kinds of media products produced by ordinary tourists. In addition to viewing traditional photo albums, we viewed digital photographs, slideshows, edited holiday videos and home pages in order to examine how such products are stored and circulated. We also explored how travel narratives and snapshots were displayed, communicated and distributed on the internet, by mail, MMS (Multimedia Message Service) or on personal home pages, blogs and other Web 2.0 platforms such as Flickr or Facebook.

Conclusion

The recent embracing of performance and non-representational theory in tourism studies has produced illuminating accounts of the multiple doings and enactments at play in modern tourism. However, in this chapter we have argued that this performance turn has not so far developed a full methodological repertoire capable of grasping the many mobilities that produce and are tied into the staging and performing of tourism. In spite of its innovative conceptual work, the performance turn has been trapped by 'the local' and by *single*-sited ethnographies. Limiting our attention to a priori scripted tourist places does not produce new knowledge about the role of tourism and travel in contemporary societies, but contains the danger of continuing a blind reproduction of received knowledge of what tourism is and should be about.

To remedy this, we have argued that studies of tourism performances can be constructively developed by broadening the scope of the methods and sites that are employed and researched. This chapter has stressed how multisited methods and mobile methods can be helpful in this regard, and we have coined the notion of mobile tourism ethnographies. Drawing inspiration from multisited research, it has been argued that tourist researchers should expand the range of research sites to include not only tourist places (resort areas, attractions and so on), but also the flow spaces of transport and the spaces of private homes. This, we argue, will allow us better to capture the network-like character of tourism performances, which is the ambition of the present book.

Enlarging the scope of the particular sites chosen for study in this way and enriching our methodological creativity with the way we engage with them will, we argue, not only produce illuminating accounts of tourism performances, but also, and more generally, make us understand better the role of tourism within the broader context of modern everyday life. By highlighting networks, paths, chains and threads between a complex multiplicity of sites, rather than examining in detail performances and practices within single sites, such an approach acknowledges the network-like character of how contemporary cultural geographies are produced, networked and performed.

4 Material cultures of tourism

Introduction

In Chapter 1 we noted the need to make issues of materiality more central to discussions of tourism performances, which is the aim of the present chapter. Despite the fact that tourists constantly interact corporeally with umbrellas, walking boots, sunglasses, sunbeds, benches, walking paths, beaches, souvenirs, maps, suitcases, cars, cameras and many other things, and physical places, tourist studies have failed to understand the significance of materiality and objects in modern tourism, the 'sensuous immediacy' of material culture to tourists and performances of tourism. Like much theory and research influenced by the 'cultural turn' in the social sciences, tourist (and leisure) studies has melted 'everything that was solid into air', or, even better, signs (Jackson, 2000; Philo, 2000). By emphasizing cognitive and human processes such as thinking, imagining, interpreting and representing, it has dematerialized bodies, things and places as culturally inscribed signs or imagescapes, to sign-value. Such a perspective has been blind to the fact that non-human things, such as objects and technologies, enable human agency and are crucial in making leisure and tourism geographies happen-able and performable. It has wrongly portrayed the world of leisure and tourism as a purely human accomplishment.

This chapter examines recent writings within sociology, geography and anthropology on material culture and hybrid geographies with an agenda of promoting non-representational and hybrid theory in leisure and tourist studies (Hinchliffe, 1996, 2003; Murdoch, 1996; Dant, 1998, 1999; Miller, 1998; Whatmore, 1999, 2002; Ingold, 2000a; Michael, 2000; Crouch, 2002, 2003a). This literature shows how objects, technologies and material environments can no longer be evaded by social and cultural theory because culture and social life is intricately tied up with and enabled by various non-human agents. It is argued that the material, cultural and social are not autonomous worlds, but intertwine and interact in all kinds of promiscuous combinations. And the inescapable hybridity of 'human' and 'non-human' worlds is stressed (Thrift, 1996: 24).

This work on hybrid geographies and material culture has recently inspired tourism researchers to write accounts of tourism as embodied, multisensuous and technologized performances through which people are actively involved in the world, imaginatively *and* physically (Löfgren, 1999; Crouch, 2002, 2003a,b; Franklin, 2003; Obrador-Pons, 2003; Bærenholdt *et al.*, 2004; Duim, 2007a,b). Such studies enable us to see how tourist performances involve, and are made possible and pleasurable by, objects, machines and technologies. We argue that tourism things are crucial in tourism performances primarily because they have *use-value* that enhances the physicality of the body and enables it to do things and sense realities that would otherwise be beyond its capabilities. Tourism is much more tied up with physical sensations than traditionally assumed.

We begin by briefly discussing how cultural studies in tourist studies have been dominated by a visual or representational paradigm that has examined material culture in relation to symbolic value through semiotic analysis with the result that the use-value of things has been trivialized. Then we examine how recent writing on material culture and non-representational theory enables us to materialize tourist studies without killing the imaginative and emotional qualities of tourist encounters. We then materialize and embody the spaces and places of tourism. It is shown that a practice-orientated performance perspective can illuminate the heterogeneous and enacted thing-ness of landscapes as well as tourists' hybrid performances *within* and *upon* them. The last part of the chapter is more empirical in nature. We begin with a general discussion of *corporeal* mobility as a hybrid performance. Then, based upon our ethnographic fieldwork from Turkey and Egypt, we discuss how the material cultures of swimming pools and bazaars afford specific corporeal performances and sensuous geographies.

Materializing tourism cultures

The hegemonic position of the representational in cultural studies of tourism illustrates the dematerialized nature of much tourist writing. In MacCannell's writing on authenticity (1976/1999), Urry's (early) notion of the tourist gaze (1990, 1995), Shields's work on 'place myths' (1991), Selwyn's edited collection on 'myths and myth making' (1996) and, indeed, Hitchcock and Teague's (2000) book on souvenirs, material objects exist as symbolic entities that humans see, photograph and engage with mentally, but never touch or interact with bodily. Natural surroundings and objects are seen as signifying social constructs that can be unveiled through authoritative semiotic readings rather than in terms of how they are used and lived with in practice. As a consequence, numerous studies have examined how postcards, brochures, paintings and other representational material instruct tourists' gazes and inscribe places with fantasies and power relations (e.g. Goss, 1993; Dann, 1996a,b; Edwards, 1996; Selwyn, 1996; Markwick, 2001; Waitt and Head, 2002). This has produced insightful accounts of the symbol values of tourism objects, but rendered insignificant their use-values. This partly reflects the fact that, in the social sciences, culture is conventionally treated as something mental and human, a 'way of life' without thing-ness, occupying the minds of people and their social representations. Such accounts separate humans from non-humans to create a genuine 'social world'. In so doing they produce an artificial dualism between culture and materiality in which the former dictates the latter by inscribing discursive worlds into and upon them. To quote Ingold:

> Understood as a realm of discourse, meaning and value inhabiting the collective consciousness, culture is conceived to hover over the material world but not to permeate it. In this view, in short, culture and materials *do not mix*; rather, culture wraps itself around the universe of material things, shaping and transforming their outward surfaces without ever penetrating their interiority.
>
> (Ingold, 2000b: 53)

Instead of positioning humans over and against the material world, Ingold suggests a shift of perspective. Drawing on Merleau-Ponty, Ingold (2000a,b) argues that the human body is not so much in space as belonging to space. Bodily practices are already orientated towards actions in the world (Merleau-Ponty, 1962: 142). Hence, humans are inscribed in the world and do not merely ascribe meaning to it. They inhabit it from their birth onwards, they use it, and their capabilities (language, the ability to use tools and so forth) are products of this active use. This argument is in line with Latour's assertion that things and humans do not exist without one being full of the other. To consider humans necessarily involves considering objects, and vice versa. In this sense, non-humans 'empower' humans and enable 'agency' (Latour, 1993, 2000; see also Michael, 2000; Whatmore, 2002):

> Things might authorize, allow, afford, encourage, permit, suggest, influence, block, render possible, forbid and so on. Actor network theory is not the empty claim that objects do things instead of human actors: it simply says that no science of the social can even begin if the question of who and what participates in the action is not first of all thoroughly explored, even through it might mean letting elements in, which for lack of a better term, we could call *non-humans*.
>
> (Latour, 2005: 963)

From such perspectives 'culture exists neither in our minds, nor does it exist independently in the world around us, but rather is an emergent property of the relationship between persons and things' (Graves-Brown, 2000: 4). Culture is a *relational* achievement between humans and non-humans, 'involving the creative presence of organic beings, technological devices and discursive codes, as well as people, in the fabrics of everyday living' (Whatmore, 1999: 26; see also Bingham, 1996: 647). It follows that we can no longer equate culture with intentionality and linguistic competences because this will exclude the work of objects and reproduce a mistaken dichotomy between language and the world (Whatmore, 1999: 30). Discourses, sensuous bodies, machines, objects, animals and places are choreographed *together* and build heterogeneous cultural orders that have the capacity to act, to have effects and affects. Material cultures are simultaneously practical, expressive and symbolic; they are heterogeneous (Michael, 2000), having both sign-value and use-value. The relationality between things and people in material worlds is clued up with imaginations, cultural styles, feelings and emotions (Sheller, 2004). 'In a culture which favours bricolage, simulation, performativity and acting-as-if, we have increasingly learned to calculate and play with this radical indeterminacy between the real, the not-so-real and the imaginary' (Pels *et al.*, 2002: 3). Following this work, we suggest that it is necessary to leave behind 'the tourist as such and to focus upon the contingent networked performance and production of places that are to be toured and remade as they are toured' (Bærenholdt *et al.*, 2004: 151). Franklin (2003: 98) argues similarly that what ought to interest us in tourism is precisely the '*links* and *relationships* between humans, machines, animals and plants and an enormous universe and variety of objects' and the effects they produce.

By moving from discursive models towards more corporeal and object-mediated ones, we can stress how tourists also encounter things through the hands, through corporeal proximity as well as distanced contemplation. We become *involved* with things. From this perspective, tourists engage with material cultures because they are *useful*. Things and technologies can be understood as 'prostheses' that enhance the physicality of the body and enable it to do things and sense realities that would otherwise be beyond its capability (Parrinello, 2001: 210; see also Lury, 1998). Tourist things acquire value – *use*-value – through being employed in embodied and poetic practices, in and through the sensuous materiality of the body. The dramaturgical landscapes of tourism's material culture comprise physical places, fantasylands and media worlds in a single human world of possibilities, of experiments with identities, social roles and relations, and interactions with places (Löfgren, 1999: 7).

The materiality of tourist landscapes

In tourist studies, discussions of place have been associated with visual consumption (M. Andrews, 1989; Ousby, 1990; Urry, 1995) and landscape as a 'way of seeing' (Cosgrove, 2003). Discussions have focused upon how landscapes dematerialize nature by transforming it into socially constructed 'wallpaper' consumed – controlled and possessed – by an aloof 'mobile eye'. The focus has been upon the 'semiological realization of space' (Ringer, 1998), or how representational cultures make space (M. Andrews, 1989; Shields, 1991; Urry, 1995). Edensor sums up:

> All too often in geographical accounts of place and space, and in accounts of human experience and practice, materiality is ignored, implying that subjective understandings emerge out of broader discursive and representational epistemologies . . . Seemingly there is no sense that embodied subjects physically *interact* with space and objects, gaining sensory experiences that shape an apprehension as to their feeling and meaning. It is therefore essential to reinstate the affordances of place and space, those qualities which are spatial potentialities, constraining and enabling a range of actions.
>
> (Edensor, 2006: 30)

By contrast, writers such as Michael (2000), Franklin (2002), Hinchliffe (2003) and Ingold (2000a) propose a 'material semiotics' or 'dwelling approach' to landscape in which material, social and cultural aspects of place sedimentation are integrated. As Cresswell (2003) and Michael (2000) argue, landscape as a 'way of seeing' presupposes an outworn distinction between humans and environments that precludes us from seeing the intimate, sensuous performances between humans and material affordances (on the concept of 'affordances' see J. J. Gibson, 1977, and Chapter 1). There is much more to the construction of landscapes than social construction: it is a heterogeneous process in which static and mobile non-humans as well as embodied, sensuous humans play their part. Thus, instead of portraying landscapes as a purely cognitive matter of inscribing already existing surfaces with beauty, narratives and myths, as tourist researchers have preferred to do, they are concerned with how landscapes are habitually and practically built up from within by the mutual involvement of humans and non-humans already dwelling and performing in the world, working from within the world, not upon it (Ingold, 2000b: 68). As Szersynski *et al.* (2003: 4) suggest: 'Out of this mutual improvisation one loses a sense of nature as prefigured and merely being "played out"; instead, the performance of nature appears as a process open to improvisation, creativity and emergence, embracing the human and the non-human'.

Such work allows Crouch to argue that tourism performances of and within nature take place in and through multidimensional spaces: 'We live places not only culturally, but bodily' (Crouch *et al.*, 2001: 259). Tourists necessarily bring their gendered, aged, sexed and racialized bodies with them, and they inescapably see, hear, smell, touch and taste the landscapes they travel in and through. Places and experiences are physically *and* poetically grasped and mediated through the sensuous body. This is the basic corporeality underlying all sensuous experiences of tourism. It is through our bodies-in-motion that we perform and 'make sense' – physically, semiotically and poetically – of spaces and places. Rather than being there simply for observation, nature is mobilized into multiple possibilities of significance. As Crouch argues in his studies of allotment practices, 'the materiality of nature may be practiced in relation to artefacts through which memory, identity and a sense of being in the world may be forged' (Crouch, 2003b: 25). The art of allotments is, in Crouch's account, in no way prefigured. Instead, creativity and texture emerge out of the repetitive performances of gardening. Nature, landscape and leisure spaces emerge from the material 'lay geographies' performed by their practitioners. They are not prefigured but made – and made sense of – through *practical* actions. Once the classic Cartesian dualism of mind and body is rejected, it follows that imagination, fantasy and 'making sense' are embodied, and part of embodiment:

> [T]he essential character of space in tourism practice is its combination of the material and the metaphorical. Once we acknowledge the subject as embodied and tourism as a practice it is evident that our body does encounter space in its materiality; concrete components that effectively surround our body are literally 'felt'. However, that space and its contents are also apprehended

imaginatively, in series and combinations of signs. Furthermore, those signs are constructed through our own engagement, imaginative engagement, and are embodied through our encounter in space and with space.

(Crouch, 2002: 208)

Although cherished for their serenity and untouched, scenic natures in tourism discourses, landscapes have a dense materiality of roads, cars, buses, bridges, buildings, restaurants, paths, viewing stations, monuments, cornfields, woods, and so on, that afford certain landscape performances and not others. Places and landscapes are encountered not 'naked' but through the deployment of a variety of prosthetic objects and technologies. Technologies are central to how people appear to grasp the world and make sense of it. They are crucial to how places are (or can be) encountered and perceived. Technologies afford and affect subsequent affordances (Michael, 2000; Ingold and Kurttila, 2001; Sheller and Urry, 2004). They are material surfaces that afford increased bodily capabilities, and as such they expand the affordances that nature permits the otherwise 'pure' body. In his fascinating study of 'mundane hybrids', Michael (2000) brings out the crucial role that walking boots play in affording leisurely country walks. They afford more pleasant walking *and* they make certain surfaces walkable that would be painful, if not impossible, to traverse barefooted or even in ordinary shoes.

When one reads historical accounts of nineteenth-century picturesque land-scape as a 'way of seeing' (M. Andrews, 1989; Ousby, 1990; Löfgren, 1999) in such a perspective, it becomes evident that this 'visual landscape' came into being only through a relational network mediated by a specialized visual sense that was based on the *technologies* of the *camera obscura*; Claude glasses and later photo-graphic cameras; *techniques* and *performances* of sketching, picturing and gazing contemplatively, with reference to visual art; *images* and *texts*, mobile drawings, paintings, photos, guidebooks, road maps and so on; and *material environments* of idyllic villages, serene landscapes and developed environments with vantage points (viewing-stations, balconies, etc.) and walking paths. This landscape vision depended on various objects and mundane technologies, and it undercut 'simple dichotomies of what is natural and unnatural, what is countryside and what is urban, and what are subjects and what are supposedly objects' (Macnaghten and Urry, 2001: 2). Although culturally constituted, it was not without a material real-ity: it circulated in mobile cultural objects; it became built into the environment; and embodied landscape performances took place in and had an effect on it. The social construction of landscape 'entails, at a minimum, the circulation of paper and bodies and manifold other materials' (Michael, 2000: 50). It is crucial to understand that landscape representations are travelling objects at once informa-tional and material (digital):

in this sense, landscape representations become dynamic vehicles for the circulation of place through space and time . . . Like Latour's scientific cir-culating references, landscape-objects allow us to 'pack the world in a box' and move about it, contributing to the shaping of the knowledge of the world itself.

(della Dora, 2007: 293)

Thanks to the travelling of images, landscapes are on the move and connected to other places and they can be consumed at a distance through imaginative travel.

Such circulating objects also afford memories of places. The individual and collective memory work of reflection and recall tends to be organized around material objects of various sorts: clothes, furniture, artwork, jewellery, personal photographs, and so on (Radley, 1990: 57–8). Although souvenirs and other memory objects, such as photographs, are often portrayed as purely semiotic (see, for instance, Hitchcock and Teague, 2000), the materiality of souvenirs and photographs is integral to their functioning, and touristic remembering is often a hybrid performance. Material culture plays a crucial role in enabling and 'storing' human memories, often in unpredictable and unconscious ways. As Marcel Proust reflected more generally:

> The past is hidden somewhere in the realm . . . beyond the reach of the intel-lect, in some material object (in the sensation which that material object will give us) of which we have no inkling. And it depends on chance whether or not we come upon this object before we must die.
>
> (quoted in Kwint, 1999: 2)

Wilson (1992) summarizes how photography permits humans to assume own-ership of nature and places as graspable, mobile objects: 'the snapshot transforms the resistant aspect of nature into something familiar and intimate, something we can hold in our hands and memories. In this way, the camera allows us some control over the visual environments of our culture' (p. 122). The magic of tour-ist photography is the way it creates immobility in an era of 'liquid modernity' (Bauman, 2000). The flux and flow of tourist experiences are '(re)solidified', ripped out of time, into something that people can hold in their hands (we elabo-rate on the materiality of souvenirs and photographs in Chapter 8).

Fragile places

In *Performing Tourist Places* (Bærenholdt *et al.*, 2004), such an alternative material reading of the landscapes and places of tourist performance is offered through the metaphor of 'the sandcastle'. Like sandcastles, tourist places are tangible yet fragile constructions, hybrids of mind and matter, imagination and presence, culture and nature. The castle comes into existence only by drawing together particular objects, mobilities and hybrid performances. There are *imagi-native mobilities* circulating in the global flows of representational objects such as photographs, postcards, brochures, films and representational technologies such as the internet and television sets, which during long winter nights help people to dream of and relive sun-drenched summer beaches. There are *corporeal*

mobilities, such as the journey to a holiday region, a day trip and the dense choreography of a family moving around and building the sandcastle. These necessitate
road networks, widespread access to private cars, holiday housing, camping sites,
beach hotels, restaurants, public holiday legislation, planning legislation, road
maps, guidebooks, ideologies of domesticity, the nuclear family, and so on. Then
there is the fine-grained sand with its alternating wet and dry textures. Then there
are *object mobilities* such as dead fish, stones and mussels by the shore or on the
beach that may have travelled thousands of miles and are gathered together by
eager children's hands and helpful parents, fetching buckets of seawater for the
moat. The tools for building, such as buckets and spades, are brought in the family
car (perhaps from a neighbouring country) but will have been manufactured in
and transported from China or a similar low-wage country. Buckets and spades
enable the mobilization of sand and water that is necessary for human hands to
build the sandcastle.

The sandcastle is a heterogeneous order, drawing together imaginative, corporeal and object mobilities. These mobilities are all elements of the networks that
stabilize and regulate the sedimented practices that transform the endless mass of
white, golden, fine-grained sand into a habitat: a kingdom of sand imbued with
dreams, hopes and pride (Bærenholdt *et al.*, 2004: 2; see also 49–68 on the beach
as leisure landscape). The sandcastle illustrates how places are dynamic, 'places
of movement' (Hetherington, 1997; Coleman and Crang, 2002). They travel, slow
or fast, over greater or shorter distances, within networks of humans, objects,
discourses, technologies and material environments. Places are about relationships, about the placing of people, materials, images and the systems of difference
and similarity that they perform. Places to tour are themselves toured by touring
actors, objects and imaginative geographies materialized and mobilized in and
through photographs, films, television programmes, souvenirs, clothes, food, and
so on (Bærenholdt and Haldrup, 2004; Sheller and Urry, 2004). This landscape
is what Ingold would call a *taskscape* (2000a; and see Edensor, 2006). With the
notion of taskscape, Ingold (2000a: 196) refers to the ways that humans routinely
inscribe themselves in space by using, inhabiting and moving through it: 'Just as
the landscape is an array of related features so – by analogy – the taskscape is of
related activities'.

The notion of taskscape highlights how people interact corporally, multi-
sensously, routinely and creatively with landscapes. They step into the 'landscape
picture' and they engage bodily, sensuously and expressively with their material
affordances. Throughout this engagement they build landscapes and places, such
as the sandcastle. Places to inhabit and dwell in.

The sandcastle further demonstrates how the existence of a particular physical environment, object or technology does not itself produce a tourist place;
these are nothing but potential – dreams of something that may happen. A pile of
appropriately textured sand and sandcastle tools is nothing until there is embodied
activity, performances of nature. Obrador-Pons (2007) argues in a similar way for
a non-representational material reading of the nudist beach as a tourist landscape.
The sensuous geography of the beach is primarily a haptic one:

The beach is first and foremost a haptic geography. Being naked on the beach is a matter of feeling rather than seeing. The direct and often sensual contact with the environment, the feeling of the sun caressing the skin, the sensual movement of the naked body into the seawater and the unpleasant infiltration of sand into bodily orifices; all suggest a haptic order of the sensible.

(Obrador-Pons, 2007: 134)

Leisure practices most often involve a sensual involvement with such material landscapes, their affordances and the pleasant and unpleasant effects (pain, fear, joy, excitement) they give rise to. This is not only limited to leisure landscapes such as the beach or the woods, in which the haptic geographies of resting, touching or walking connote an idea of landscape as a 'milieu of corporeal immersion' rather than 'a visionary moment of drama' (Wylie, 2005: 242). In addition, places and landscapes prefigured as iconic places of drama and vision are contingently produced as hybrid, floating places. Considering the emblematic symbols of the Orient, the pyramids of Giza outside Cairo in Egypt, Mitch Rose proposes that:

how a landscape matters (how it has material effects on our lives) is directly connected to how it matters (how it comes to be significant within a network of meanings and relations. Thus the physical being of landscape, its ongoing presence in the world is contingent upon what it initiates, activates and inspires elsewhere.

(M. Rose, 2002: 456–7)

He furthermore demonstrates how the myriad of interests and performances unfolding in and around the plateau of Giza continuously (but contingently) translate these 'piles of stone' in Cairo's suburb into a tangible material *and* symbolic place: although it appears as a definable material space, its materiality is 'constituted by the totality of possible performances immanent within it' (ibid.: 463). Thus, the particular leisure landscape of the Giza plateau is produced through a multiplicity of mobilities, networks and performances affording particular haptic and visual geographies to be performed (see Chapter 6 for more on this). Although the pyramids may impress the observer by their overwhelming presence, the plateau has to be contingently produced and performed to emerge as a 'place for tourists'. Thus, the places, sites and landscapes of leisure and tourist performances such as photographing, socializing or shopping are not what simply is 'present-at-hand', there to be encountered by the disinterested subject. Rather, they are 'ready-to-hand', already involved with the everydayness of the embodied subject, making them appear in the world through practices of interpretation and incorporation. Drawing loosely upon our ethnographic fieldwork, we now briefly flesh out the embodied and sensuous geographies emerging as part of three emblematic material landscapes for mundane tourist performances (sights, swimming pools and bazaars).

Sights, pools and bazaars

Sights

The task of ordering the spatial flows of sight-seers at tourist attractions is often delegated to material objects such as signs, paths, benches and, not least, viewing platforms (Figure 4.1). At most historical sites, omnipresent signs and clearly demarcated paths instruct tourists how and where they are supposed to walk and 'sight-see', and they may forbid inappropriate performances such as touching, climbing and playing on ruins. Their job is to *prevent* unnecessary and potentially harmful *physical* contact between tourist bodies and the precious object of the tourist gaze and to secure safe and smooth walking and guiding to the major viewing stations.

At the castle of Alanya, these viewing stations turn the castle into a wonderful view-producing machine. They are viewing stations where tourists can consume laid-out panoramic and imposing sights. As a towering fortress it has had a powerful 'military gaze' looking out for enemies built into it from the start: today it is tourists' cameras that do the shooting. Walking is put on hold, eyes gaze, fingers press shutters, bodies pose and loud travel talk is heard: 'Wow! What a wonderful view!' The viewing platforms and tourists' cameras turn their backs on the castle itself, and they face the 'cliffscape' and Alanya city. The castle of Alanya is not so much an object of the 'tourist gaze' as it is a machine producing bird-like views

Figure 4.1 The castle in Alanya.

of the sea and city. This reflects the fact that the tourist gaze is not only a way of seeing and representing landscape, but also a way of making landscape material so that it presents itself as a nicely framed picture.

Pools

Another major icon of package tourism is the swimming pool, and our ethnography shows how many tourists – especially families with young children – spend much of their time there on a daily basis. And yet the swimming pool remains thoroughly under-researched. One reason for this is that tourism studies have largely reduced the significance of tourism to gazing and ignored the bodily pleasures of tourism. Another, but related, reason is that tourist studies have adopted an elitist view on tourism:

> It is difficult to avoid the suspension that most writers have generalised about tourism by accounting for the touristic dispositions and experiences of a small group of middle-class Western tourists in particular kinds of tourist space and have neglected, for instance, working class-class seaside holidays and carnivalesque adventure.
>
> (Edensor, 2006: 26)

It is sad that Franklin and Crang can state so correctly that the tourist literature reveals very few clues that 'pleasure, fun and enjoyment' are central features of tourism (2001: 14). Swimming pools are places of bodily 'pleasure, fun and enjoyment', and we may add relaxation. Much like the nudist beach discussed above, the sensuous geography of the swimming pool is a haptic one: of being touched by the warm sun and jumping into the refreshing water. But there is much more to the swimming pool than the sun, the water and seminaked bodies. Without a whole array of both static and circulating objects, the swimming pool would probably not be a place of much 'pleasure, fun and enjoyment', or perform-able or inhabitable. The most significant objects are pictured in Figure 4.2. They include sunbeds, plastic balls, books, magazines, waterslides, airbeds, sunshades, swimming wings for children and inflatable playthings. These objects are constantly in use and inhabited. One dwells in a swimming pool by engaging with such material objects. The adjustable plastic sun chairs offer both sleeping and reading or 'gazing' positions; the sunshades provide crucial breaks from the heating sun; light books and magazines afford relaxing reading; various plastic toys and the waterslide make it possible to have fun and play in the water. Perhaps the most significant and used object of the swimming pool is the airbed. This object is used not on land but in the water. Not unlike a surfboard (discussed below), the airbed affords a 'supernatural' experience of lying on the refreshing water and yet one is still touched by the tanning sun. Lying on its cool texture, with one's hands and feet dipping into the water, one can be in burning sun for long time. And crucially, one can always roll over into the refreshing water!

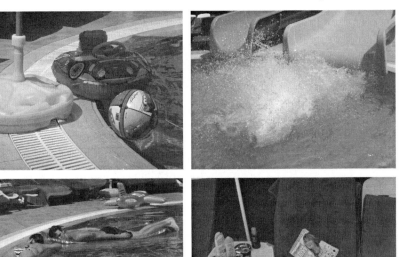

Figure 4.2 The swimming pool.

Bazaars

Accounts of tourist performances often fail to see mundane performances of walking and strolling in the urban landscape of towns, cities and holiday resorts. Yet strolling around in such landscapes often occupies plenty of tourists' time. A study of package tourists to Benidorm, Spain, found that people spend almost two hours a day promenading the urban landscape of this particular resort (Iribas, 2000: 112–13). Navigating, drifting through and even inhabiting promenades, alleys, shops, restaurants and cafés of tourist places are major practices through which tourists make themselves at home through interpreting and incorporating the material spaces of their holiday. Often the very culturally specific and different ways of organizing the urban space of inner cities is both a major obstacle for tourists as well as a major way of familiarizing themselves with unknown and unexpected people, places and cultures. Thus, Rodaway (1994: 57–60) suggests that the inner environs of Arab cities, dominated by alleys, courtyards, souqs and so on, provide a much congested and immediate haptic geography, a geography that affords close proximity for transactions as well as a 'cellular' structure unfamiliar from Western urban design. Thus, the stroll through the bazaar streets in the souq is perhaps first and foremost an encounter with a visual, olfactory and auditory geography alien to most tourists. Mastering the souq with its – to most Westerners – iconographic yet mute road directions given in Arabic is an exercise in collective map-reading and way-finding through the labyrinth of alleys and crowded main streets. In the labyrinthine streets of Cairo's Khan el Khalili bazaar one's body is constantly brought into contact with other bodies moving busily here and there, smells of spices, rotten fruit and other people's odour and sounds of crying shopkeepers advertising their goods (Figure 4.3). Holding on to their purses, children and bags, tourists may try to negotiate good bargains, but talking, chatting and bargaining are not easy for the visitor unfamiliar with the multilayered sensuous geography of the souk.

Then the grid-like bazaar of Sharm el Sheikh's 'Old Market' affords a much more familiar space for negotiating what constitute proper souvenirs, appropriate gifts or good bargains (Figure 4.4). Shopping in the bazaar is not just about purchasing goods. It is also a way of sensing, dwelling and making oneself at home, of connecting the social and the material, by collectively exploring and bargaining as well as bringing the prey back home. In the bazaar, dwelling is performed through a series of encounters with the sensuous and material landscape of the city and purchasing global brands such as Swatch, Nike or Lacoste or trivialized iconographic souvenirs such as glass pyramids and toy camels or pieces of craft to be exhibited on the shelves of the lounge at home or passed on to friend and family members as gifts (for more on this, see Chapter 8).

Figure 4.3 Khan el Khalili, Cairo.

Figure 4.4 Mobilizing the camel.

Movement as a hybrid performance

In the previous sections we have discussed places and landscapes as material sites contingently produced by a variety of mobilities in space. Now we want to add one further aspect to this discussion of mobility and place, namely the role played by 'sensed mobility' in making particular sensuous geographies performable. Many tourist practices are intriguingly tied up with performances of mobility such as kite-surfing, water-skiing, bicycling, and so on (see also Chapter 6 for discussions of the role of camel-riding in performing tourism).

Hybridized mobility is such an obvious characteristic of tourism (and almost all other aspects of the social world) that it has been partly overlooked until recently. In tourist studies it has largely been reduced to a precondition for performing tourism, a practical issue of 'getting there' and 'getting around' rather than a way of sensing movement and landscapes. However, cars and other 'mobility technologies' are central to how people grasp, inhabit, sense and make sense of landscapes and places, and they are central to how tourists perform places and derive pleasure, enjoyment and experiences from them. Haldrup (2004) shows how driving a car on holiday enacts very different 'modes of mobility' with regard to families sight-seeing from at home. Thus, family life and leisure places are simultaneously produced through the hybridized performance of driving a car.

Other recent writings within social theory on 'automobility' have similarly suggested that we should understand driving and passengering as hybrid per-formances 'in which the identity of person and car kinaesthetically intertwine'

(Thrift, 2004a: 47; see also Katz, 2000; Urry, 2000; Miller, 2001b; Sheller, 2004). The embodied and sensuous experience of movement is kinaesthetically sensed through our joints, muscles, tendons, and so on, as we move in and across the physical world. The 'automobilized person' is simultaneously inhabiting and feeling the car, but he/she is feeling the physical world through the moving car, as it moves. The hybrid performances of the 'driver–car' (Dant, 2004) afford specific physical and virtual effects and sensuous experiences. The hybridized performances of movement generate 'supernatural' sensuous effects. Mobility hybrids signify freedom, transgression, excitement and speed.

The very performance produces such effects. First, while trains, cars and buses impoverish the other senses, they greatly enhance and alter visual sensations. They are prosthetic vision machines that interject their own spatio-temporal reality, a cinematic experience of place (Schivelbusch, 1979: 58; Larsen, 2001). Framed by the cinematically mobile windscreen, the tourist's visual perceptions multiply, and become dynamic and chaotic: 'a rapid glance in magic glass' (Retzinger, 1998: 216). Second, hybridized movement in and through spaces creates kinaesthetic effects and sensations that would otherwise be beyond human experience. In the early years, cars and trains were looked upon as shocking speed machines. Schivelbusch summarizes (1979: 58): 'the train [was] experienced as a projectile, and travelling on it, as being shot through the landscape – thus losing control of one's senses'. Decades later, the speeding potency of the car caused similar excitement and disdain: 'the exultation of the dreamer, the drunkard, a thousand times purified and magnified' (Liniado, 1996: 12; see also Sachs, 1992).

Another example of this is the surfboard. Making a surfboard perform-able requires detailed sense-derived knowledge of local wave conditions, knowledge often shared and communicated within particular social groups, magazines, and so on (Preston-Whyte, 2002; Shields, 2004). It also requires extensive embodied and practical knowledge of what can be done with the board – or, if one is windsurfing, also of mast, sail and boom. Such knowledge is 'meaning embodied – feel, touch, fluid – and possibly not speakable?' (Game, 1991: 57). 'When windsurfing is successful', as Dant (1999) says, 'it is unclear to what extent the human is responsible and to what extent the windsurfing equipment is' (p. 128). The same can be said of driving, skiing, paragliding, rock-climbing, sailing, and so on. Dant nicely brings out the magical excitement that the hybrid body–board–wave assemblage provides:

> It can enable a human being to skim over water, blown by the wind, in a manner at once natural and yet so alien to the body. The usual limitations of the human body keep it immersed in the water, vulnerable to being buffeted by the wind wherever it is exposed. But the object of the windsurfer enables its sailor to transcend these constraints, raising the person onto the surface of the water and harnessing the propulsion of the wind. The resulting experience of the speed, with a certain acrobatic grace, liberates the body from its humdrum uses and experiences. . . . The role that the wind-sailor adopts is mimetic of the experience and excitement of a form of mobility in space

that would normally be the preserve of birds or mythical beings – skimming over or walking on the water. In their rubber garb, harnessed to the rig and with feet linked to the board through toe-straps, sailors are transformed like mythical characters from their ordinary state of being.

(Dant, 1999: 112, 118)

The sensuous experience of being a mobile hybrid is 'at once natural and yet so alien to the body'. The 'windsurfer' invigorates the human body, makes it active, absorbed and responsive to the collective rhythms and fluids of body, windsurfer, waves and winds, with little (wet jacket) or no bodily protection or shielding. These 'natural' interactions produce the supernatural experience of skimming and flying over the water, somewhere between a fish and bird: this hybrid performance transports the tourist far away from his/her 'ordinary state of being. And much the same can be said about other mobilities of 'adventure tourism' such as river-rafting, bungee-jumping, rock-climbing, skiing: they too involve material objects, hybrid performances and material interactions with landscapes. Such tourism has a history. In 1932, Tom Stephonson organised trips to the Peak District, England, 'not to see landscape, so much as to experience it physically – to walk it, climb it or cycle it' (Cosgrove, 1984: 268). Adventure tourists are physically engaging with nature's physical powers. Landscapes are energies. Material surfaces and fluids are experienced directly through the active, moving, hybridized body. They are touched with all the senses and pulsating rushes of adrenaline runs through the body: 'Nature, for many tourist consumers, has evolved from something to look at, to something to leap into, jet boat through, or turn completely upside down it: the inverted sublime!' (Bell and Lyall, 2002: 27).

Conclusion

Tourist (and leisure) studies, as well as sociology and cultural geography, have shared a common view of the social world in which objects, things and technologies are too easily and quickly disregarded or seen as alienating intruders. We have argued that objects are almost inescapable parts of most tourism performances. Tourism is not a flight from the world of things. In tourism, people interact routinely with a wide range of objects and material environments; they bring their gendered, racialized and aged bodies into play when performing leisure and tourism. And we have argued for a material reading of tourist *bodies* by showing that we need to move beyond the tourist gaze and explore sensuous geographies of doing, movement, touch and smell. Against this background, we have suggested that tourism's material cultures should be uncovered, thus bringing out objects, technologies and material environments and showing that these are simultaneously material, cultural and social, place-bound and mobile. In other words, we have shown that cultures are heterogeneous.

In this chapter we have argued for the necessity of moving beyond the sign-values of 'tourist' things and tourist places, engaging with their use-values and affordance. We have shown how things work in daily leisure and tourist practices when sensuous bodies and responsive technologies are combined to create various 'supernatural' effects and particular tourist and leisure spaces and places. Hybrid performances such as those discussed in this chapter both have effects and produce affects (see Thrift, 2004b). Tourism and leisure research will continue to be blind to the significance of things, materialities and technologies if it does not engage with the various effects generated by hybrid performances such as sight-seeing, driving, and so on. Such performances have important material effects on how places and landscapes become physically structured. Moreover, hybridized performances also generate affects and emotions. Things, technologies and the physical layout of landscapes constitute a thick materiality of emotions, long-ings and thrills that will not be uncovered unless both representational and non-representational geographies are accounted for. In particular, we have shown how significant tourist places such as swimming pools and bazaars in part 'seduce' tourists because of their material geographies that afford various bodily pleasures that involve geographies of touch, smell and movement as much as gazing.

Thus, this chapter does not advocate neglecting the cultural and representa-tional aspects of leisure and tourist performances. Instead we argue that tourist researchers should strive to be 'more-than-representational' (Lorimer, 2005). We need to uncover the role of apparently trivial objects and technologies in human agency. Challenging unimaginative accounts of material culture that reduce things and objects to soulless matter, we propose to bridge the gap between, rather than separating, the material and the immaterial, the concrete and the metaphorical, the dreamed-of and the lived-in orders of reality. Such an approach to 'materiality', we argue, does not reinforce an essentialist dichotomy between the 'objective' (material) and the 'subjective' (textual); rather leisure and tourist studies must work with a complex notion of culture that privileges neither the material nor the immaterial. This is a question not of returning to the material, but of bridging the gap between what has been separated. This is the path we follow in the rest of this book.

5 Mobilizing the Orient

Introduction

In Chapter 1 we noted the strange similarities between the parallel and convergent evolution of hegemonic paradigms in cultural accounts of Orientalism and tourism. At a closer look this 'strangeness' may not be so strange. Both privilege ideas of cultures as geographically rooted, visual and textual readings of culture and (re)produce a dichotomous view on the 'everyday' and the 'exotic'. In Chapters 2–4 we developed in more detail how we might mobilize and materialize studies of tourism and, by doing this, uncover the everydayness that lies at the root of most tourist endeavours. In this chapter, we take this one step further by fleshing out how 'the Orient' is invoked in various tourist and non-tourist contexts and circulated in different material and embodied forms through performances of contemporary everyday lives. In doing this, tourist encounters with choreographed 'Oriental' tourist places in Turkey and Egypt are framed by experiences and expectations generated by the various materializations and mobilizations of 'the Orient' in the media, in the streets and on computer screens 'at home'.

In Chapter 6 we track the practices and movements of tourists through the fabricated spaces of the Orient in the streets and hotels of Egypt and Turkey. However, the relations that produce the leisure spaces of the Orient are not bounded to the spaces of tourist resorts and attractions. In the present chapter we show how material mobilities prefigure and frame encounters with 'the Orient' in contemporary everyday lives. This chapter then serves two purposes. First, we 'mobilize' and 'materialize' the concept of Orientalism. Second, we show how Orientalism makes its way into contemporary everyday lives by showing how Orientalist practices frame the home context for (Danish) tourist travel to the Middle East.

Travel and Orientalism are (and historically have been) closely related. Taking our starting point in studies of travel literature of the Orient, we discuss how we may think of Orientalism in terms of performance, embodiment, materiality and mobility rather than (only) imagination and representation. We then show how embodied encounters with 'Oriental' others are pre-formed by embodied and material encounters 'at home' by discussing three examples of encounters with the Orient-at-home. All three examples capture how the Orient in various ways

is mobilized, materialized and circulated and through them pre-forms performances in – and pre-frames encounters with – the people, places and cultures of
Turkey and Egypt. 'Oriental places', the first example, is the so-called 'cartoon
controversy' that took place when the research for this book was enacted. We
examine the global circulation of cartoons that were originally aimed at Danish
newspaper debates on migration and identity politics as an example of how material mobilities in unexpected ways prefigure encounters between the Orient and
the West. The second example shows how discourses of Orientalism are translated
into everyday practices and habitual spaces of ordinary experience, and how 'cultural things', such as the cartoons, effectively draw on the embodied encounters
and affects. In encounters with migrants, refugees and strangers, Orientalism is
mobilized and enacted in people's home geographies when borders between 'us'
(Westerners) and 'them' (Oriental others) are routinely established and maintained. Third, and finally, we discuss how tourists using the wide range of media
technologies, through their own representational practices, circulate Orientalist
tropes thereby prefiguring encounters with the Orient as part of contemporary
cybercultures (this is further developed in Chapters 7 and 8).

Travel and the cultures of Orientalism

It has been repeatedly shown how Orientalist tropes frame tourist encounters and
imaginations. Central to this line of research is the idea that the hegemonic discourses of the tourist industry – through the recirculation of stereotyped images in
postcards, souvenirs and tourist brochures – reproduce a particular geographical
imagination of the Middle East (Edwards, 1992; Ryan, 1997; Gosden and Knowles,
2001; Burns, 2004; Bryce, 2007). In doing this, they show how Orientalist representations circulate through centuries-long 'vicious' circles of interpretation
and still emerge as stereotyped and banal elements in tourist discourse. However,
such studies seem to be stuck within a representational framework that tends to
separate the 'cultural' from the material and the performative. The practices of
Orientalism are part (now more than ever!) of material and more-than-representational practices that produce, frame and anticipate tourist encounters with distant
Others in Turkey, Egypt and beyond. The glossy pictures and seductive accounts
of the Orient in travel brochures, films, literature and art are not limited to a world
of reflections and representations. Instead, as Gregory (2005) suggests, we must
acknowledge the substantial reality of the Orient's 'imaginative geography'.

Following Said's work (1995), Mitchell's *Colonizing Egypt* (1988) shows how
this iterative and citationary character of early tourists' travel accounts of the
Orient were closely related to:

> the colonial process [that] would try and re-order Egypt to appear as a world
> enframed. Egypt was to be ordered up as something object-like. In other
> words it was to be made picture-like and legible, rendered available to politi-
> cal and economic calculation.
>
> (Mitchell, 1988: 33)

Although the detailed registration, enframing and displaying of the customs, manners, artefacts and architecture of ordinary Egyptians astonished, amused and embarrassed contemporary Egyptians (see Mitchell, 1988, Chapter 1, for examples), they were essential for Europe's colonial relation with the Orient. By photographing, sketching, collecting, exhibiting and narrating Europe's Other, 'the Orient' was ordered as a visual and consumable object. Mitchell also describes how travellers and tourists arriving to Egypt in the late nineteenth century inevitably tried to grasp the Orient as though it was an exhibition of itself (Mitchell, 1988: 28–9). Searching for 'the real Orient' beyond the representations on the Great Exhibitions or in contemporary travel literature they inevitably discovered that:

> The Orient was something one only ever rediscovered. To be grasped representationally, as the picture of something, it was inevitably to be grasped as the reoccurrence of a picture one had seen before, as a map one had already carried in one's head, as the reiteration of an earlier description.
>
> (ibid.: 30)

As a contemporary reviewer of travel literature complains in 1852: '[t]here they are; the same Arabs, camels, deserts, tombs and jackals that we journeyed with, rode on, traversed, dived into, and cursed respectively, only a week ago' (cited in Mitchell, 1988: 30).

Writing about late nineteenth-century visitors to Cairo, Gregory (1999, 2005) shows how the imaginative geographies that came out of travel writing, literature about and guidebooks to the Arab world were not only crucial for how Cairo was perceived and portrayed but also 'instrumental in orchestrating their [the tourists'] performance in – and *of* – Cairo' (Gregory, 2005: 93). European tourists travelling to Egypt consumed Cairo as an Orientalist space of fantasy in which it was made visible as the timeless city of the Arabian Nights. Cairo became a place of real fantasies where the space of the image and the space of the city were constantly folded into one another. One tourist wrote: 'Egypt is an open-air museum where temples and tombs are arranged like shop windows for public inspection' (cited in Gregory, 1999: 134). Sailing up the Nile in 1948, another one wrote: 'One had to rub one's eyes to be sure that one was not in the theatre . . . [it] was like a sublimated opera scene' (cited in Gregory, 1999: 115).

The Egypt encountered by these tourists was produced by searching out elevated positions and open vistas or by sailing down the Nile in a *dahabeeah* (a large luxury houseboat with cross-sails) and by looking through the camera. Distance was essential to order and eradicate Cairo's turmoil. And inability to achieve distance for a good 'shot' at the unfolding spectacles and awe-provoking monuments was a recurrent source of annoyance. As one glancing Kodak-wearing *dahabeeah* passenger wrote: 'A vision of half-barbarous life passes before you all day and you survey it all in the intervals of French cooking: Rural Egypt at Kodak range – and you sitting in a long chair to look at it' (cited in Gregory, 1999: 131).

This reiterative process of representation and performance were further enhanced with the organization of package tourism (including Nile Cruises) by Thomas Cook & Sons in 1869 and the launching of user-friendly, lightweight cameras by Kodak in the 1880s. These innovations in transport and communication technology enabled 'ordinary' (though mainly well-off) tourists and travellers to Egypt to participate in the circle of representing, circulating and performing the Orient.

But the Orient was not only circulated on film and in literature. Tourist performances of the Orient rest in significant ways on the inscription of material landscapes through architecture and urban design with Oriental styles and iconographies (such as the pyramids) – material elements that can also be mobilized and materialized to different degrees and in different geographical settings to invoke leisurely performances of the Orient. Mitchell (1988) especially emphasizes the significance of the mid-nineteenth-century Great Exhibitions in Paris and London for circulating the Orient. Other examples include themed environments such as the world-famous Tivoli Gardens in Copenhagen (which explicitly borrowed the citation of 'Oriental' styles – from Chinese to Arab! – in its architectural layout from the Great Exhibitions), the Pharaonic Village on the outskirts of Cairo (Slyomovics, 1989) or the famous icon of post-modernity, the Luxor Las Vegas Hotel (Cass, 2004).

In Chapter 1 we argued that the theatrical metaphors of the performance turn should be stripped from its connotations of falseness and make-believe and that the notion of performance may help us in grasping practices rather than texts: the social construction of reality rather than its representation. With this in mind, we see that Said's metaphor of the Orient as 'a theatrical staged affixed to Europe' does not just capture how 'the Orient' is a show (or 'exhibition') put on for the gaze of the Western viewer (Said, 1995: 1, 63). Instead, we may conceive of the Orient as materially fabricated by networks of humans and non-human objects affording (or restricting) particular performances. Not only are performances 'scripted' by Orientalist discourse but the stage materializes such discourses into set-pieces, artefacts and roles. This fabrication of spaces of tourism performances is a laborious process involving the materialization of Orientalist literature, art and architecture in the streets of the cities as well as the eradication and removal of unpleasant inhabitants and their everyday places. Through such relational material interventions and mobilizations of a host of technologies, artefacts, material spaces, the production and ordering of 'the Orient' takes place, as performance rather than as picturing.

Orientalism as performance

The etymological roots of 'Orientalism' are the same as in the verb 'to orient'. Both are to be found in the Latin *oriens* (rising, east). This common etymology dates back to the common way of 'orienting' maps to the East (Holy Land) before the widespread use of technologies such as the compass, which 'reoriented' cartographic imaginations to the magnetic North Pole. The double meaning of the 'Orient' as both a noun (a geographical direction) and a verb (to be able to navigate) also encapsulates the function of Orientalism in Said's seminal work. To Said, 'Orientalism' was never pure representation. As he points out throughout his work, the acts of representations that produce 'Orientalism' in fiction, journalism, drama, music, science, and so on, are not innocent 'cultural' or symbolic expressions idealizing, romanticizing and covering a 'real' world to be uncovered 'underneath' such 'cultural' layers. Rather, these acts of representation are powerful steps in exercising power over and producing 'the Orient' *as* a (material, embodied) reality.

Said, and much of the work inspired by him, effectively demonstrates the hegemonic discourses involved in producing and reproducing colonial domination by the West over the East. Orientalist discourse was (and is) basically, in Bryan Turner's words, 'a discourse which represents the exotic, erotic, strange Orient as a comprehensible, intelligible phenomenon within a network of categories, tables and concepts by which the Orient is simultaneously defined and controlled' (Turner, 1994: 21).

As Mitchell (1988) shows, the discourse of Orientalism was decisive in reordering the Middle East as subject to colonial exploitation. Thus, the power of representation implicit in Orientalist discourse and its role in exercising colonial domination was produced by a huge machinery of artistic work, government reports, travel literature, scientific dissertations, and so on. In doing this, 'Orientalism offered . . . not just a technical knowledge of Oriental languages, religious beliefs and methods of government, but a series of absolute differences in which the Oriental could be understood as the negative side of the European (Mitchell, 1988: 166). However this division of the superior West and its hegemony over the subjugated East was not only produced as a work of knowledge and discourse, a projection of the imagination and categorization of artists, scientists and imperialists from European (and North American) centres of art and knowledge. It was also established through a process in which:

> railways, steamships, telegraphs, newspaper correspondents, official reports, photographers, artists, and postcards from the front were all being brought into coordination. . . . In this manner the enormous truth of colonialism – both its description and its justification – could be ordered up, put into circulation and consumed.
>
> (ibid.: 168)

But the (re)production and circulation of Orientalism were not only a product of a material machinery. The *effects* of Orientalism are also material and embodied. Through banal acts of representation, Orientalism establishes the very difference between the East and the West, and transforms the former into 'materials' for the imagination of the latter. As Grosrichard (1998) shows, the Orient emerges in Western culture as an amorphic, dark chaos onto which the repressed fantasies of a rationalized 'enlightened' society are projected. The Orient always emerges within an ambivalent geography providing both demonized, frightful spaces to be conquered, confronted and controlled and inviting, pleasurable and seductive scenes and creatures to be consumed.

Writing about eighteenth- and nineteenth-century accounts of travels in the Middle East, Kabbani (1994) emphasizes that:

> to write a literature of travel cannot but imply a colonial relationship. The claim is that one travels to learn, but really, one travels to exercise power over land, women, and peoples. It is a commonplace of Orientalism that the West knows more about the East than the East knows about itself; this implies a predetermined discourse, however, which limits and in many ways victimizes the Western observer.
>
> (ibid.: 10)

Kabbani, however, argues that it is not in the role as an ideological language or narrative legitimizing Western domination over the East that the true significance of Orientalism is to be found. Rather, it is by producing and providing an intriguing theatrical space upon which Western fantasies can be enacted and staged. She argues that Orientalist discourse provides a script in which:

> The Orient becomes a pretext for self-dramatisation and differentness; it is the malleable theatrical space in which can be played out the egocentric fantasies of Romanticism. It affords endless material for the imagination, and endless potential for the Occident itself.
>
> (Kabbani, 1994: 11; see also Behdad, 1994; Hutnyk, 1996)

In this way, the equation of verb (*to* orient) and noun (*the* Orient) becomes clear. The discourse of Orientalism provides the plastic material by which the Orient emerges as a material and theatrical space, which in turn enables Orientalist fantasies and imaginations to emerge and be played out. 'Orientalism' is basically a 'spatial story' or, in Said's own words, an 'imaginative geography'. This geography is, however, not a 'cultural' thing, but basically a device that *produces space*.

This is also emphasized by Gregory (1999, 2004), who points out that a main element in Said's work is the recognition that 'distance, like difference – is not an absolute, fixed and given, but is set in motion and made meaningful through cultural practices' (Gregory, 2004: 18). 'Orientalism' is in itself performative as it names the effect it produces. Thus, Gregory argues that the citational structure of Orientalism:

> is in some substantial sense performative. It produces the effect that it names. Its categories, codes and conventions shape the practices of those who draw upon it, actively constituting its object (most obviously "the Orient") in such a way that this structure is as much a *repertoire* as an archive . . . [I]maginative geographies are not only accumulations of time, sedimentations of successive histories; they are also *performances of space*. . . . [S]pace is an effect of prac-

tices of representation, valorization, and articulation; it is fabricated through and in these practices and is thus not only a domain but also a 'doing'.

(ibid.: 18–19)

Following Gregory we may think of Orientalism in terms of a repertoire of techniques, technologies and practices that enable (geographical) distance and (cultural) difference to be enacted and performed. Drawing in particular on the work of Mitchell (1988), Gregory emphasizes that the powers of Orientalism were based not so much in its representational aspects, by providing a representation of the spaces of the Middle East, but rather in its capability to *enframe* these spaces. In other words, in Mitchell and Gregory's perspective, the point is *not* that Orientalism provides a (distorted) representation of the world, but that it is a tool for making sense of the world and exercising control over it. Its 'imaginative geography' is about 'doing' rather than 'viewing', and the substantial significance of Orientalism is then that it produces material and embodied effects and affects. The 'substantial reality' of Orientalism rests on its performative powers: its ability to produce the effects it names.

As a repertoire of technologies, tools and practices for establishing order (difference/sameness, distance/closeness) and exercising control, Orientalism has been mobilized and materialized in a variety of geographical settings beyond the Middle East and the Far East (paradoxically absent in Said's work!). Orientalist techniques migrated to the south and the west and proved to be powerful tools in the encounters with 'other' places, people and cultures on the African continent (Merrington, 2001; Kasfir, 2004) and the Caribbean islands (Sheller, 2003; Tzanelli, 2007: 121).

Writing about the latter, Sheller (2003: 144) observes that the consumption of the Caribbean 'occurs first through its displacement from the narrative of Western modernity (decontextualization), followed by its recontextualization as an Other to serve the purposes of Western Fantasy'. From the first Western encounters with the West Indies, explorers, travellers and colonizers described and depicted these in the mirror image of the 'East Indies', scripting the former as an 'Asia of the West' (Sheller, 2004: 121). However, Sheller shows how this 'Orientalism' was not accidental. By transporting the Orientalist vocabularium to the West Indies, these well-known tools and practices of colonial domination were mobilized and brought to work within a new context of exploitation and domination. By following the simple Orientalist formula quoted above, Caribbean consumer products, fruits, vegetables, places, landscapes, villages and bodies were inscribed with distinctively 'Orientalist' features – and consumed:

> It is not only things or commodities that are consumed but also entire natures, landscapes, cultures, visual representations, and even human bodies . . . [T]here are crucial forms of consumption 'at a distance' which must also be considered. . . . It is not only 'goods' which circulated in the transatlantic world economy, but also people, texts, images, desires, and attachments. To bring into focus the full range of consumption linkages between 'advanced'

consumer societies and the post-slavery societies of the Carribean, it is necessary to foreground the forms of mobility which connect here and there, then
and now . . . Playing on the interlocking meanings of 'Carib' and 'cannibal'
ever since Columbus' confused arrival in the New World a typology of forms
of material and symbolic consumption can be proposed. These include inges-
tion, invasion, incorporation, infection, appropriation, sacrifice, and exhibition, as well as various processes of possessing, destroying, using up, and
wasting away.

(Sheller, 2003: 14)

Stressing that this consumption of the Caribbean is an *active* practice that produce particular social relations and roles, Sheller shows how the techniques and
practices of Orientalism are mobilized and moulded to the particular socio-spatial
context of the Caribbean islands. Like Mitchell (1988) and Kabbani (1994),
Sheller emphasizes how Orientalism serves as a powerful tool or technique for
transforming and making 'other' people, places and cultures into objects of domination, control and consumption by enacting material mobilities connecting 'here'
and 'there'. Orientalist techniques are not only crucial in making 'other' spaces
control-*able* and manage-*able,* but also in making these spaces and the cultures,
places, objects, goods and people belonging to them consume-*able.*

Here it may be constructive to consider the idea of performance again. In
Chapter 1 we observed a general move from semiotic interpretations of the symbolic values of consumer goods and tourist places towards material readings of
practices and performances of consumption and tourism. We furthermore argued
that this move implies a sensibility to the expressive and creative capabilities of
users-as-performers. In doing this we may acknowledge consumers to be knowledgeable, creative agents rather than passive recipients of symbols and goods.
Thus, redirecting attention from the hegemonic discourses (re)produced through
literature, science, art, media and so on to multiple performances of Orientalism
we may uncover the practical everyday expressions, uses of and responses to
Orientalist techniques and practices that are mobilized in consumption and everyday life. In the remainder of this chapter we discuss three such examples of how
Orientalism is performed in encounters between East and West.

Practical Orientalism

The 'cartoon controversy'

One of the most telling examples of how Orientalism mobilizes everyday life
in unexpected and surprising ways is the so-called 'cartoon controversy', which
took off following the publication of 12 cartoons of the prophet Muhammad in a
regional Danish newspaper (*Jyllandsposten*) in the autumn of 2005.[1] The afterlife
of the 'cartoon controversy' took the form of recurrent violence (across Muslim
countries in the Middle East and Asia) and symbolic enactments of cultural difference (in Denmark and Europe more generally).

First, the basic course of events. In the late summer and early autumn of 2005 it was rumoured in the Danish press that an author of a children's book, Kaare Bluitgen, had had difficulties finding an illustrator for his new book on the life of Muhammad. In public debate this issue was assumed to be related to 'similar' incidents in Europe. Following the murder of the Dutch filmmaker Theo van Gogh (2004), newspaper editorials and politicians from the Danish right-wing coalition soon called for 'a defence for the right of free expression' in the face of Muslim aggression in Europe. As part of the public debate that followed, the major regional newspaper *Jyllandsposten* decided to invite cartoonists affiliated with the newspaper to draw caricatures of the prophet. On 30 September 2005, 12 drawings were published under the heading 'The Face of Mohammad'. Not all of these were explicitly derogative towards the prophet, or Muslims in general. But three drawings, one depicting a sinister-looking religious leader with a bomb in his turban, one depicting 'Mohammad' with a sword and one showing a Muslim jihadist being calmed by his leader, clearly drew on a demonizing Orientalist imaginary concerning threatening 'fanatic' Muslims. These three drawings particularly provoked citizens of Muslim background and many others.

At first the drawings received little attention outside Denmark. On 17 October, the Egyptian newspaper *El Fagr* reprinted six of the drawings with a critical commentary yet without stirring much debate. Meanwhile, in Denmark, events escalated as spokesmen for Muslim communities interpreted the drawings as the latest in a series of xenophobic injustices aimed at non-European minorities in Denmark (see next section). This they reported to Muslim and Arab embassies in Copenhagen, and on 12 October 11 of these ambassadors wrote to the Danish prime minister to discuss what they perceived as an 'ongoing smear campaign' against Muslim minorities. And from this point on the cartoon controversy took on a life of its own. The official response to the letter was that the press in Denmark enjoys the right of free expression and that the government was not inclined to interfere with this right and thus that there was no need to discuss anything. When this reply was reported in the Middle East press and copies of the drawings began to circulate (together with other examples of anti-Muslim statements – including an interview with the Dutch filmmaker, author and member of parliament Ayaan Hirsi Ali), the cartoons became an issue of global controversy. In February 2006 Danish embassies were attacked by demonstrators in Beirut, Damascus and Tehran (in Denmark a simulacrum of these events took place when a school was burned down by a group of 'second-generation' immigrants) and global media circulated pictures of burning Danish flags on the Palestine West Bank, the hanging of dummies named after the Danish prime minister in Pakistan, and so on. Now the drawings became topics of global discourse, and Western political leaders (most prominently the former US president Bill Clinton) criticized the drawings. Eventually the case calmed down although further 'riots' occurred occasionally in Denmark and the Middle East, e.g. in September 2006 (the anniversary of their publication) and in 2008 (following police investigations in Denmark into an alleged plot to kill the cartoonist behind the 'Bomb in the Turban' drawing).

Three stories can be told about this chain of events. First, the official response by government and media press representatives in Denmark was that the 'cartoon crisis' reflected an ongoing lack of understanding by the Muslim communities of the inviolable right to free expression in democratic societies such as Denmark (Hansen, 2006; Ammitzbøll and Vidino, 2007). Thus, the events were basically a clash of values and civilizations, with the modern, secularized and enlightened West, on the one side, and the traditional, religious and parochial East, on the other. In that sense, the 'cartoon controversy' is part of the more general struggle between the West and the East, of which the US-led Global War on Terror is also a part.

Second, this Orientalist interpretation has been challenged by 'multicultural-ists,' who argue that this controversy represented a struggle for recognition for all the involved parties. They argue that lack of recognition of the collective religious identity of Muslim minorities in Denmark rather than the actual content of the drawings was at the centre of the conflict. The 'cartoon crisis' reflected an emerg-ing split in Danish society caused by the poor integration of its non-European citizens (Bleich, 2006; Lægaard, 2007). Whereas the 'Orientalist' interpretation stresses the needs of the East to comply with Western civilization, the 'multi-culturalist' interpretation points to the need for mutual understanding across cultural differences.

Slavoj Žižek proposes a third interpretation. For Žižek, the controversy reflects how global mobilities draw together distant sites and places in one unified space so that people are constantly – and uncomfortably – confronted with habits of thought and doing that they do not understand and dislike. 'It is as if Denmark and Syria, Pakistan, Egypt, Iraq, Lebanon and Indonesia really *were* neighbour-ing countries' (Žižek, 2008: 50). The imperative of understanding across cultures therefore has to be complemented by an attitude of 'getting out of each other's way', what Žižek calls 'a new "code of discretion" ' (ibid.: 50).

It is a striking feature of the cartoon controversy that the participants in it were talking at cross-purposes and were unable to understand each other's motives and intentions. It also seemed that they did not always know what they were really talking about or, for that matter, responding to. In the Middle East, the drawings were first circulated by discussion groups on the internet and through text messages, with the result that only fragments and rumours found their way from Denmark to the Middle East (Bonde, 2007). And when reported in the Arab media, the cartoons were presented as part of the ongoing 'transgression' by the West towards Muslim cultures. Thus, the general role of Arab media in the initial stages of the controversy was to legitimize anti-Western sentiments and actions in general rather than addressing the content (Douai, 2007). Most of the protesters on the streets in the Middle East had not seen the drawings, nor did they have a clear picture of the incidents triggering their violence. On the other side of the circuit, the burning embassies and flags, 'rioting mobs' and broadcasted terror threats against Danes (and Europeans more generally) simply reinforced an Orientalist imaginative geography of a Middle East made up of burning war zones and popu-lated by fanatical, unfriendly and rioting mobs. It seems that little knowledge or understanding was transported through the networks and flows of global media: no 'rational' strategic warfare was fought, but a lot of anger, fear and injury were created. Thus, the 'cartoon controversy' is comparable to a cultural car crash – a contingent and chaotic effect of the (in this case) malfunctioning machinery of global mobilities.

Encountering embodied others

As outlined above, the 'origin' of the cartoon controversy was in part to be found in the parochial ambivalences of globalization in Denmark. The incident followed soon after leading Danish politicians had stated their discontent with migration regulations aimed at immigrants from Muslim countries in statements such as 'by 2005 human beings at a lower level of civilization . . . populate big parts of Copenhagen and other major Danish cities with their foreign, cruel habits' (leader of the Danish People's Party) and calls from the Minister for Culture for a cultural battle against Muslims not complying with Danish culture (reported in Bonde, 2007: 36). Such expressions were the immediate pretext of the 'cartoons controversy', and they reflect an official recognition by politicians in power of what some scholars have seen as a latent but invisible cultural racism emerging in Denmark in the 1980s and 1990s (Wren, 2001: 146). In recent years, discriminatory accounts of, and acts towards, especially refugees and immigrants from Muslim countries became part of discourses of banal nationalism and practices of institutions and large proportion of the (ethnic Danish) population, although the process was 'subtle and almost invisible' in the beginning (ibid.: 146, 158). Yet the calls for 'cultural battle' against Muslims and the 'cartoon controversy' reflect how the once 'invisible' processes of demonization and 'Orientalization' of migrant cultures in Denmark surface and become visible in political discourse and public debate.

As Said also has frequently argued, Orientalism has no ontological stability. It has to be continuously produced and reproduced. The distinction produced between the West and the Orient is a supreme fabrication: 'lending itself to collective passion that has never been more evident than in our time with mobilization of fear, hatred, disgust and a resurgence in self-pride and arrogance (Said, 2004: 870).

In other words, the theatrical choreography of Orientalism has to be continuously staged and performed, arousing affects of pleasure and repulsion to remain stable (or change). Such everyday performances of difference may be coined 'practical Orientalism', as opposed to a purely discursive and representational conception of Orientalism. With this wording we want to stress the significance of everyday consumptions and (re)negotiations for reproducing Orientalism as a natural, self-evident, 'taken-for-granted' moral order and emphasize that:

> Orientalism is not . . . passively reflected in everyday life – it is rather distributed, manipulated, reproduced and opposed. It is linking the little banal social

poetic with the grand dramas where contrasting images between the Orient and the West are fought in real visible wars, exclusions and repressions.

(Haldrup *et al.*, 2006: 175)

These Oriental dramas of everyday life are clearly expressed in the accounts of everyday life collected in Denmark in the early years of this century. Thus, it is striking that, when discussing encounters with multiethnicity in everyday life, (some) people would draw up an imaginative geography of confrontation and domination (Haldrup *et al.*, 2006, 2008). Moreover, these geographies are profoundly sensuous, reflecting how they emerge as part of people's practical bodily involvement with the world (Rodaway, 1994; see also Chapter 3). One of the respondents quoted in Haldrup *et al.* (2006), for example, pictured a haptic geography of physical violence and corporeal struggles in the streets of Copenhagen in which youngsters of Muslim origin excluded socially marginalized ethnic Danes. According to this person 'alcoholics and drug addicts are a threatened species . . . by the second generation immigrants. . . . And they [second-generation immigrants] clearly think that you can just hit them and kick them and do whatever you want to do with them' (ibid.: 180). Another complained about being forced into an auditory geography at the work place in which Arabian and Turkish languages dominated the soundscapes. To this person, the very sound of Arab and Turkish languages was sensed as 'psychological racism against the Danes . . . We are after all in Denmark, aren't we? And then they are sitting there cackling in their own language' (ibid.: 182). This person explicitly feared being mocked behind her back. Her account, however, reflects a more general sensuous and imaginative geography in which Danes/Europeans/Westerners are marginalized by intruding Orientals. Other examples of how this practical Orientalism is enacted and performed through the establishment of sensuous geographies of smells, taste, food, sounds, embodied poses and visuality relate to topics such as school catering and veiled teenage girls. They all reflect the consolidation of a strange and reverted imaginative geography in which demonizing and intruding Orientals conquer and occupy the everyday spaces of ordinary (ethnic European) people. In this way, the dominant Westerner is strangely victimized by the dominated.

Here we have provided only one example of how the subtle and invisible workings of practical Orientalism contribute to the consolidation and production of difference through material and embodied performances. As we shall see in Chapters 6 and 9, the embodied and habitual responses described above are also part of the techniques and practices people draw on when performing tourism. They are, so to speak, part of the tourist's baggage; the 'other' is encountered, experienced and interpreted through the lenses that tourists bring with them from home. As H. Andrews (2005) shows, the habitual reproduction of 'banal nationalism' (Billig, 1997) is a significant aspect of the way people make themselves feel at home in tourist resorts (see also Chapter 2). Towels, clothes, tattoos, food and drinks 'flag' identities and borders are drawn within the tourism resort. This again highlights how the construction of 'difference' is not a purely symbolic process but also a profoundly embodied performance:

[I]t is how tourists *feel* that is important to them rather than what they see. The point is that the textual surface of Union Jacks, English-language menus, television programmes and 'British' foods is visually incorporated into the body to give rise to feelings, understanding and thus knowledge of what it means for these tourists to be British . . . Thus the tourists in this context do not 'gaze upon sights of difference' but smell and hear difference.

(H. Andrews, 2005: 263)

Establishing borders between home and away, the well-known and the exotic, is both a source of enjoyment and annoyance to tourists on holiday. This ambiguity is constantly in play in sensuous and embodied encounters with foods, living environments, embodied social norms, and so on. At the same time, it is also a significant way that people make sense of the world they are confronted with and how they frame, perform, re-present and circulate their own personal and sensuous experience of the Orient that we will now discuss in the third case.

Circulating the Orient

We now return to the issue of representation and circulation in relation to performances of the Orient. As we have seen, imaginative geographies of the Orient are established through iterative processes of performance *and* representation, a process in which embodied and material encounters and practices form the non-representational machinery producing and circulating the Orient. The representational practices involved in depicting and describing other (Oriental) places, people and cultures are not innocent. As Gregory observes with reference to the current 'Global War against Terror', 'people go to war because of how they see, perceive, picture and speak of others: that is, how they construct the difference of others as well as the sameness of themselves through representations' (Gregory, 2004: 20).

It is Orientalist fantasies about dehumanized and demonized Arab Muslims that generate the affects that in turn enable their humiliation and torture by American soldiers in Abu Ghraib and elsewhere in Iraq (Gregory, 2007: 228). But, as we see in the case of the thousands of photographs taken by American soldiers in Abu Ghraib, the production of Orientalist imageries presupposes the production of a very material and embodied space in which such violence may take place:

torture requires its victims to be less than human, so that the degradation can continue – the spectre of 'the monster' stalks the torture chamber – but it also requires them to be human . . . A space that is once inside and outside the politico-juridical order is a space in where these double subjects can be conjured in to being, paraded and subjugated.

(ibid.: 229)

In the prison cells of Abu Ghraib:

the camera literally worked as weapon of war (as an instrument of torture). The denial of shared humanity, the killing of the other as a human person, is a prominent feature in all these photographs. The victim is not seen as a person with a face and a history.

(Laustsen, 2008: 130)

The immense ideological power of photography is its seductive realism, which conceals photographs' constructed nature. In contrast to painting, sculpture or writing, the ontological power of photography is its magical claim to open up a window to a real, pre-existing 'reality' without human inferences: photography does not distort or shape the world but extends it, uncovers it, mirrors it and captures it on film or as digital bits in a computer chip. When photographs are understood to be taken rather than being *produced* they do not allow for the same interpretive flexibility as other representational technologies: they simply document (Sontag, 2004a: 41). 'Here I am and this is what (really) happened.' Thus, photography's non-human aspects entirely overpower the human ones: cameras 'do' photography.

This realism is what horrified people around the world when the photographs of triumphant soldiers staging and executing brutal humiliation and torture of Iraqi prisoners circulated in the global media in 2004. It is telling that one of the first reactions from the US State Department was to regret – not the violence, torture and humiliation committed – but the behaviour of the soldiers: that they behaved like ordinary tourists by circulating their personal snapshots on the internet and through their camera phones (reported in Laustsen, 2008). By using digital technologies, the soldiers produced and instantly circulated their personal stories of triumph and exultation to their family and peers back home at the same time as they choreographed the violent spaces of an Iraqi prison populated with half-human monsters to be subjugated by force. In this way, the photographic representation of war and torture in these pictures also reflects, as Sontag observes:

a recent shift in the use made of pictures – less objects to be saved than messages to be disseminated, circulated. A digital camera is a common possession among soldiers. Where once photographing war was the province of photojournalists, now the soldiers themselves are all photographers, recording their war, their fun, their observations of what they find picturesque, their atrocities – and swapping images among themselves and emailing them around the globe.

(Sontag, 2004b: 26)

Personal mobile phone images of fun and conquest intended for friends and family members back home took wrong turns on their journey and became public images of disgrace on a global stage.

We now turn back to the issue of 'ordinary' tourists' visual representations of the Orient. The literature on digitalized communications often emphasizes the potential significance of *instantaneous* communication. This does not imply that photographic *memories* are disempowered, but rather that photographic memories are dislocated from the personal spaces of the home to the (semi)public spaces of the internet (see Chapter 8; Larsen, 2008). Thus, digital photography and Web 2.0 have enhanced the circulation of the Orient and enabled it to exist as a prosthetic memory in the distributed networks of camera phones, emails and the worldwide web. The instructive authority of travel literature and guidebooks has been democratized and personalized as 'everyone' can publish their holiday photographs, memories and recommendations. Tourists have become reviewers, advisers and guides, contributing a new sincerity and personal authority to the review and advice people may seek before boarding a plane.

However, tourists do more than just circulate advice on particular beaches, pubs, restaurants hotels, and so on. Their exhibited photographs on the internet also circulate iconographic, stereotyped tourist scenes, landscapes, people and animals. Paraphrasing the words of Mitchell's review of travel literature (1988: 30; see above) in the mid-nineteenth century, we may note that, even though (many) people take great pains at creating interesting and aesthetic pleasing photographs, it is, still, the same snake charmers, riverbanks and village children that we journeyed with, encountered and travelled along a week – or a century – ago that are depicted (Figure 5.1).

The distribution of personal stories and photographs also allows tourists to try, like the travellers of the nineteenth century discussed earlier in this chapter, to get off the beaten track and to capture the 'real' Orient: to get behind the staged. When 'commenting' his tour 'Back to Back Streets . . .' (Figure 5.2) on the photo home page trekearth.com, the 'tourist photographer' says the following: 'continuing my quest far from the main streets . . . where you find no stalls with cheap crap for tourists, where walls do not shine, where people's smile is not bought for a few bucks spent on unneeded souvenirs'.[2] Another one writes a comment about a photograph he took of some boys he encountered in the streets of Cairo:

> for those not having English as their first language, [horse play] means 'rough play,' the kind normally associated with boys. That's what's going on here – the boy in yellow expressing dominance over the one in orange, who has jumped into the street in front of me to escape. If they were horses, they'd be rearing their front legs into the air, for sure. How odd it is when we humans claim we are somehow different from other animals.[3]

Overwhelming numbers of personal, often tourist, photographs of the Orient are available on user-generated photography and travel sites such as Virtualtourist, TripAdvisor, TrekEarth and Flickr. For instance, at Flickr there are 249,500 photographs 'geo-tagged' Cairo, 18,850 results for Alanya and 35,784 results for Sharm el Sheikh (survey done 29 December 2008). Although many of these photographs are personal, the citationary structure of tourist photography is still evident. A search for a specific tourist attraction such as the Giza Plateau outside Cairo, Egypt, on TrekEarth generates an abundance of almost identical photographs (Figure 5.3).[4]

Figure 5.1 The virtual album. http://www.christianehoej.dk Photographer Lars K. Christensen.

We have thus seen how digital photography and Web 2.0. afford the endless circulation, exchange and exhibition of iconographic photographs of Orient; photographs that are ready to be encountered, grasped and uncovered at home by virtual tourists and potential travellers. Such circulating photographic representations in

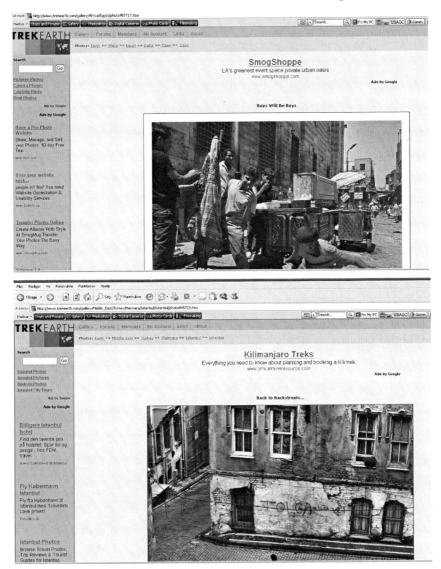

Figure 5.2 Oriental back-stages.

multiple shapes and forms both enable and restrict the repertoire of possible tourist performances. Although the spectre of Orientalism still to some extent haunts
the photographs circulated by the internet by tourists, it does not predetermine
performance. Rather it provides the techniques and tools that – together with the
material landscapes of holiday spaces – afford a variety of performances (see
Chapter 6). Representational 'products' then are not to be conceived as merely
'scripts' to be carried out. Rather the practices of representation and circulation tie

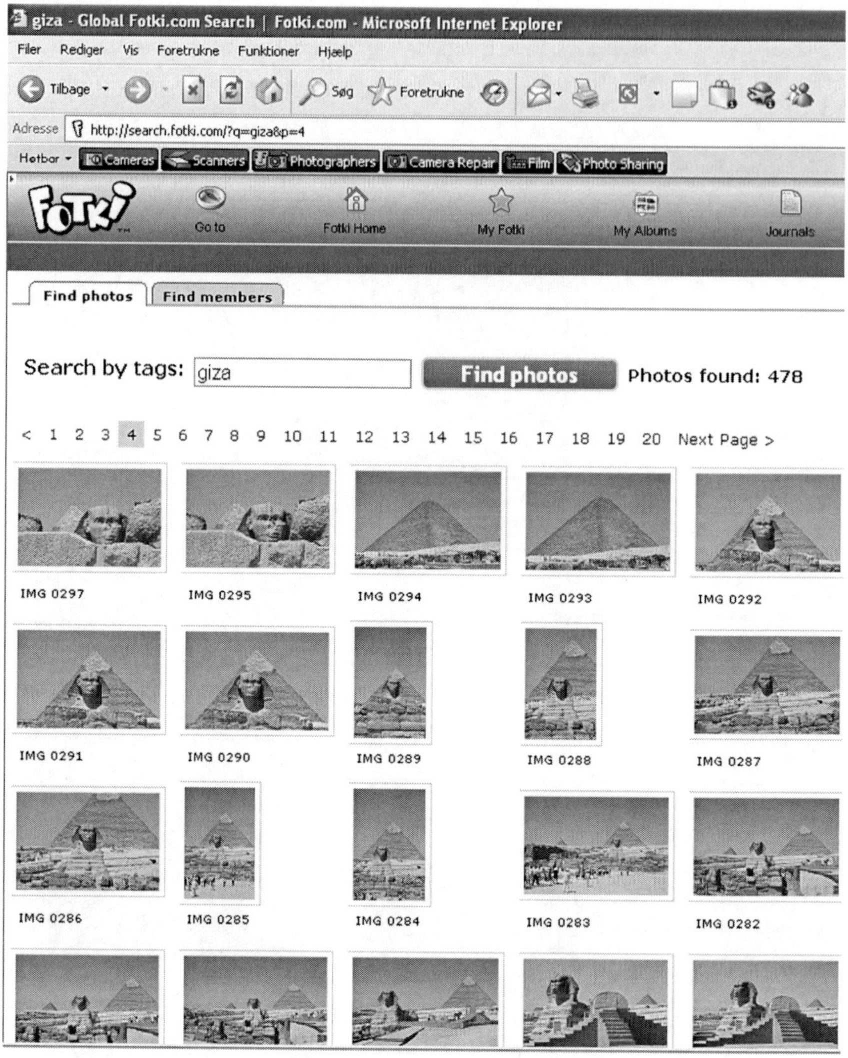

Figure 5.3 Circulating the pyramids.

playfully into performances of pleasurable holiday life, extending the knowledge, memories and performances of tourism into the fabric of the everyday 'at home'.

Conclusion

In this chapter we have seen how a 'performance' perspective on the notion of Orientalism enables us to think in terms of mobilities and materialities rather than solely discourse and representations. Drawing on the work of Mitchell (1988) and Gregory (1999, 2004, 2005), we have argued that the imaginative geographies of the Orient are (re)produced through embodied and material performances. In doing this we have shown how we may think of 'Orientalism' as a technological device that has been mobilized and materialized in a variety of geographic contexts to enable the manageability and consume-ability of 'Other' people, places and cultures for the Western subject. Redirecting the issue of Orientalism from questions of discourse and representation to practices and performance enables us to trace out the material, embodied and technological networks that constitute relations with an Oriental 'other' and enable a variety of performances of 'the Orient'. We have shown how the material circulation of cartoons through heterogeneous geographical and cultural contexts created a cultural crash in the case of the 'cartoon controversy'. We furthermore demonstrated how people's sensuous encounters and embodied experiences of 'the Other' are intrinsic for producing 'imaginative geographies' of the Orient in the home geographies of Westerners. Finally, we examined how the rich possibilities for fabricating, distributing and communicating visual representations have reinforced the circulation of the Orient and transformed the ways in which personal accounts of it are stored and communicated.

This chapter has emphasized how the Orient is circulated and makes its way into mundane and personal everyday knowledges. In the next chapter we follow in the footsteps of tourists in order to examine the many ways the Orient is performed 'in motion' on holiday travels to the Oriental spaces of Egypt and Turkey.

6   Doing tourism

Introduction

> Yesterday, ALL the significant papers in DK published one of the infamous cartoons again-again. Apparently as a kind of public gesture of solidarity with the cartoonists after the arrest of three immigrants that allegedly had planned an assault on the cartoonist behind 'The face of Mohammad' (the one with the bomb in the head). Danish media are totally self-oscillitating into hysteria about this, fearing (hoping?) for a 'cartoon controversy II'. Checked the *Al-Jazeera* website last night and they seemed well-balanced in their report about the incident. I do not hope it will interfere too much with the fieldwork – cancelled sight-seeing trips etc. At the check-in, I asked the smiling lady from the agency if any passengers had cancelled because of the re-publication of the cartoons. She was not too eager to answer – seemed distracted, probably thought it was more important to make sure that all managed to get their baggage checked in at the right place. 'No, we're full', she said, 'most likely they haven't noticed it, and those who have, don't care'. Check-in is very efficient, and we proceed to the security area where we queue in stringent lines. Waiting impatiently, unbuckling my belt. I hope that they won't ask me to take off my shoes, too. I hate standing with the shoes in my hand and trousers dropping down, humiliating! 'It's all part of the fieldwork experience', I think ironically with a nod to David Lodge's ever-energetic tourism anthropologist in *Paradise News* (Lodge, 1991: 11).
>
> (Michael's field diary, 14 February 2008)

How to write up the pains and pleasures, hopes and despairs, of 'doing' tourism? In Chapters 1 and 3, we critically addressed current uses of ethnographically inspired methods in studies of tourism and advocated a more multisited and mobile approach, which we called mobile tourism ethnographies. In Chapter 3, we argued that mobile tourism ethnographies involve more than registering the mobilities, immobilites and moorings that tie into the field of tourist performance. They also presume that the researcher actively partakes on holiday travel with (other) tourists and visits tourists at home to hear their retrospective reflections. In *Tales of the Field*, John van Maanen notes, when discussing what he terms 'impressionist tales', that fieldwork 'is a long social process of becoming. It is a process that begins before entering the field and continues long after one leaves it' (Van Maanen, 1988: 117). Ethnographers' accounts (like tourists'!) do not exclusively relate to the observed field, but are always 'mobile' in the sense that they are constructed across a variety of fields, genres and scales. In the process of writing up 'ethnographies' of people and cultures, ethnographers often have to make use of 'dramatic recall': building accounts from an assemblage of words, metaphors, phrasings, imagery and memories of field experiences. In this chapter, we let a variety of voices speak.

In order to 'track' tourists' doings to, from, within and beyond the spaces of their holiday in Turkey and Egypt, we draw on a variety of heterogeneous sources such as observation, field notes, diary entries (by tourists), photographs, informal and formal interviews, conversations en route and afterthoughts made during the home ethnographies in Denmark (see Chapter 3). Furthermore, and deliberately, this chapter is written in an impressionistic style – as a 'bricolage' or montage of 'interpreted slices, glimpses and specimens of interaction' (Denzin, 1997: 247; see also Denzin and Lincoln, 2003). And we oscillate between first-person auto-ethnographic accounts and more distanced observations of, as well as generalized reflections about, doing tourism.

So in this chapter, we follow in the footsteps of tourists. Through (covert) participant observation on two one-week package tours (Alanya, Turkey) and one two-week package tour (Sharm el Sheikh, Egypt) we track the 'doings of tourism' by examining a range of tourist scenes. After briefly addressing the paradoxes of the 'take-off', we approach the 'light Orientalism' that decorates many tourist resorts and explore how the 'staging' of Otherness may be read as an embodied, sensuous and active performance with connotations of play and make-believe. We then address the classical 'sight-seeing trip', showing how this strictly regulated and choreographed space of performance contains significant elements of site-sensing as bodies and objects meet. 'Sociality' is a very important aspect of the holiday, and we take a closer look at the more 'introvert' aspects of tourism, addressing the significance of other tourists. We then go on to examine the many objects and practices that people make use of to make themselves at home at their tourist destination. In continuation of this, we briefly address how (some) tourists 'moor' their everyday life in their favourite tourist place for a sustained period of time by purchasing houses and apartments for second-home uses or by making friends with 'locals'. Finally, we discuss how people connect to absent others at home.

Take-off

> Right on time! It all seemed very efficient and regulated in the airport to me. Yet my co-travellers are grumbling as I write this (on the plane). 'Really annoying with all this waiting and stopping!' I grab the flyer (from the agency) with instructions on how to behave when we reach Sharm and count the word 'chaotic' six times, before stopping . . .
>
> It wasn't chaotic at all, actually rather smooth and efficient, and 45 minutes after landing we were at the hotel . . . Yet (some of) my co-travellers go on

about the alleged 'chaos' and 'inefficiency', while they are queuing . . . now for the toilet . . . I wonder if this is caused by the collision between restless anticipation and the forced fixation and 'rest' in waiting lounges and plane and bus seats that travel always includes?

(Michael's field diary, 14 February 2008)

Writing about his arrival to a holiday paradise in the Caribbean, de Botton (2002: 19) describes his worries about not informing his colleagues properly about going away, pain in his temples and the pressing need for the bathroom, which makes him realise: 'a momentous but until then overlooked fact was making itself apparent: I had inadvertently brought myself with me to the Island'. Often, the imaginative travel preceding the actual bodily travel to the holiday destination is a disembodied and immaterial practice: 'At home I could concentrate on pictures . . . and ignore the complex creature in which this observation was taking place and for whom it was only a small part of a larger more multifaceted task of living' (ibid.: 21). Discovering one's bodily cargo(s) are often a source of unanticipated annoyances when travelling: hurting feet (or temples), a full bladder when queuing endlessly for one toilet on a plane or restless children on a long journey.

Yet the mobilities of tourist performances are at the same time imaginative and virtual. Tourism takes place in a 'blended geography' in which materiality and virtuality are continuously entangled. This is often in paradoxical ways, as when hurting bodies disturb and act back on the (disembodied) dreams of weightless travel to a paradise island, as de Botton describes above.

The role of 'screens' for imagining, planning and performing tourism may illustrate this blended geography (Figure 6.1). At home, television and computer screens afford the anticipation of tourist dreams and their planning. In the terminal, screens monitor the flux of planes to and from the airport to assure passengers a smooth passage to the destination of their dreams (see Figure 6.1, top). Observing these screens also provides a needed moment of disengagement from travel partners or troubles, noises and other annoyances in the departure hall before moving the body into the next line (see Figure 6.1, bottom left). Or tourists may engage with their ever-present mobile phone screen, sending text messages to relatives staying at home or receiving instructions to stay in touch while absent. On the plane, screens above your seat similarly monitor the route traversed from home to the destination. By occasionally glancing at such screens, the passenger is reassured that the plane is progressing according to plan, and the inconveniences of the flight will soon be over (see Figure 6.1, bottom right). Meanwhile screens of portable DVDs and game consoles 'immobilize' zealous and bored children in their seats. And, once the plane is 'on land', mobiles are routinely turned on and obligatory 'we have landed safely' texts may make an immediate return journey. Waiting in the hotel lobby, television screens enable bored tourists to kill time or absent themselves from their holiday destination for a brief period of time, perhaps watching their favourite football team play.

Such screens are part of networked electronic machinery that mediates and regulates travelling bodies. They promise joyful experiences, frictionless travel,

Figure 6.1 Screens.

distraction from discomforts and everyday connectivity while absent. They are indicative of how tourist travel is enabled by and performed through blended geographies of corporality and mind-travelling, materiality and virtuality as well as presence and absence. There is a networked-like character to embodied tourism performances mediated through screens.

Light Orientalism

> Passed the waiting hours in the fabulous garden of 'One Thousand and One Nights' at the Winter Palace in Aswan, discussing whether we would ever want to live in such a luxurious hotel. I was in favour of it . . . wife is against.

Again the airport refuted all warnings we had heard about chaos and inefficiency – maybe we were just lucky? Sharm is evidently a different Egypt than Aswan and Luxor – more like a kind of Arabian Las Vegas. Fortunately, the hotel is reasonably laid-back – and our rooms are in a private corner.

(diary writer, February 2007)

Together with the Turkish bath (in Turkey) and the camel ride (in Egypt, see below) 'Oriental nights' is among the regular stock-in-trade of tourist entrepreneurs in Turkey and Egypt, and references to palm trees, Bedouins, belly dancing, and so on, are popular symbolic markers mobilized to create flavours of exoticism alongside sea, sun and sand. The buzzing streets of 'enclavic' tourist spaces in Egypt and Turkey are decorated with 'Oriental' signs and symbols inviting tourists to take part in activities such as belly dancing and *shisha* smoking, for enacting 'light Orientalism'.

In Sharm el Sheikh, Michael and his family stayed in Naama Bay (the tourist honeypot of this resort area). The surroundings afford an abundance of 'Oriental'-styled restaurants (yet serving 'international' dishes such as pizzas, pastas and steaks) and open-sky cafés where one can smoke the *shisha* (brought to you by Bedouin-dressed waiters) while sitting on Oriental-style carpets and cushions. They are effectively 'camps' fenced off from the street and other cafés by plastic carpets decorated with flashing 'palm leaves' (Figure 6.2). Materials such as plastic, concrete and neon lights used for staging the 'light Orientalism' of Naama Bay may resonate more with Las Vegas than with an exotic 'Orient'.

Figure 6.2 Oriental nights.

But why are tourist resorts packed with tasteless concrete and plastic set-pieces bathed in neon light? Classic tourist theory points to alienation, believing that tourists are cheated and their quest for authenticity and proximity with the Other is 'satisfied' with reified, 'faked' and kitschy substitutes. But this does not account for the ludic and often ironic circulation of Oriental tropes, clichés and fragments at resorts. 'Welcome to Disneyland Egypt', as a newspaper salesman at our hotel ironically greeted us on the first morning. The set-pieces of 'the Oriental show' (e.g. camels, Bedouins, desert/oasis, 'Thousand and One Nights') put on in tourist resorts should not be taken at (sur)face (or rather 'symbol') value. Rather than being merely representational markers of 'the exotic', they are fragments and materials for staging improvised play and often ironic performances and representations of 'the Orient'. Moreover, the core of such performances is the embodied enactment and experience of tourists' 'others', for example Oriental Others 'performing the Bedouin' or tourists dressing up and staging themselves as Oriental Others (Figure 6.3, top left and right). Thus, the performances of the Orient for and by tourists involve engagement and play, rather than pure simulation.

Tourists are not only shown 'the Orient' but also encouraged to hear, smell, eat, dance and touch it. In Chapter 4 we noted the profoundly sensuous and embodied character of tourist performances and in Chapter 5 how 'difference' is encountered, experienced and produced in embodied and multisensous ways when performing tourism. The 'Bedouin Night Trip' – a 'must' for package tourists to Sharm el Sheikh – is illustrative thereof. In what follows we (Michael) 'shadow' a group of tourists on their journey into the Sinai Desert to experience a 'real' Bedouin night.

During the pick-up routine, participants are introduced to the theme of 'real Bedouins' as the guide expounds the fate of Bedouin culture in modern society including accusations of connections between Bedouins and terror, politics of forced settlement and labour market integration. During the journey through the Sinai Desert, the company of chitchatting Westerners (mainly families with children) slowly enters into a blended geography of fiction and fact, myth and materiality. When passing the (contemporary) settlements at the outskirts of the city, we are left with only the view of the desert and the voice of the guide. She explains that the key to understanding the life of the Bedouins is the camel:

> The camel is a fantastic animal – adapted to the life in the desert. It can carry fat for 40 days and its eyelids are like sunglasses – protecting its eyes from the wind and the sand. Its feet are designed to keep its footing in the desert sand . . . In short, the camel is, like the Bedouins, completely adapted to the quietness and tranquillity of the desert.
>
> (transcript of bus guide, Sharm el Sheikh, 16 February 2008)

'One of the finest things is the camel', Flaubert wrote in one of his letters from Egypt, 'I never tire of watching this strange beast that lurches like donkey and sways its neck like a swan. Its cry of something that I wear myself out trying to imitate – I hope to bring it back with me, but it's hard to reproduce' (Flaubert cited in de Botton, 2002: 86). To Flaubert, the camel epitomizes 'the exoticism' of the Orient and its people: their stoicism, timelessness, melancholy and sensuality.

Figure 6.3 Performing Bedouins.

> My real passion is the camel (please don't think I am joking): Nothing has a
> more singular grace than this melancholic animal. You have to see a group of
> them in the desert, when they advance in single file across the horizon, like
> soldiers, their necks stick out like those of ostriches and they keep going,
> going . . .
>
> (ibid.: 86–7)

During the bus ride, the guide elaborates on the significance of the camel: it
provides milk, meat, wool and skin; it is used as a currency, a symbol of prosper-
ity (camel saddles and harnesses are highly ornamented); and camel races are
significant events of Bedouin culture. Finally, the guide carefully instructs us how
to sit (and stay) on top of a camel back. This part of the programme unsettles

many: to leave the comfort zone of the bus and mount a camel in the company of local 'Bedouins'.

After a short drive, the bus suddenly stops on a plain between some rocks. We are told to leave the bus, and numerous children and camel drivers immediately surround us and offer us camels and to tie scarves around our heads – to avoid the sand or perhaps just for the fun of it! Once we all are safely placed on a camel, some wrapped in scarves, others trying to maintain balance in the saddle, we are dragged across the plain (see Figure 6.3, middle right and left) towards the Bedouin camp. After a 30-minute ride, we approach the 'camp' – a concrete platform with makeshift shelters, a stage and a barbecue area – and dismount the camels. Reflecting on this episode, Michael writes in his field diary:

> The experience of awkwardly balancing our bodies on top of the camel was clearly unsettling to most of us. We did not really want to move our big and clumsy bodies out of the comfortable bus. But why did we do it, then? For those who managed to get into the rhythm of the animal, balancing their body, legs crossed on the top of its 'wooly' neck, the desert transcended from being an alien surrounding . . . to a landscape that could be traversed and imaginatively inhabited. 'Hey, I am Lawrence!' an elderly guy shouted when hitting his camel to get more pace. Although I doubt he reached the meditative trance of Peter O'Toole's Lawrence (in *Lawrence of Arabia*, 1962) as he felt asleep on his mount while smoothly traversing the desert sands. But he might just have tasted the flavour of it.
>
> (Michael's field diary, 18 February 2008)

Throughout the 'Bedouin Night Trip', the encounter with the Oriental Other is framed by and performed within a sensuous geography (Rodaway, 1994; see also Chapters 4 and 5). The *haptic* experiences of 'dwelling' on the back of the camel and riding across the desert plain afford the embodied imagination of inhabiting the desert and its people. Arriving at the 'Bedouin desert camp', a concrete and steel construction supplied with shelters, carpets, cushions and low 'Bedouin-style' tables, a mint tea welcomes us and seductively wraps us in an *olfactory* geography of 'different' fragrances and flavours. Once the barbecue dinner is served and the nightfall approaches, a bonfire is lit and flutes and drums entertain with an Oriental soundscape, perfect for dancing with the Bedouins and mind-travelling while gazing into the flames (see Figure 6.3). The 'drama' of the Bedouin Night Trip then enables an embodied imagination of 'difference', a performance to take part in and enjoy actively or by 'dwelling' and engaging in imaginative travel while gazing into the flames of the Bedouin camp fire. It also encapsulates more broadly how set-pieces and accessories of 'light Orientalism' are mobilized as part of tourist performances.

First, such sensuous geographies enable tourists to take possession of 'difference'. By exposing oneself to 'different' moves, scents, flavours, sights and sounds, the tourist can enjoy and engage in 'difference' by incorporating it. As Flaubert recalled, it is the mind-blowing fireworks of sensual impressions that make this Oriental exoticism so seductive.

It is like being hurled while still asleep into the midst of a Beethoven symphony, each detail reaches out to grip you, it pinches you; and the more you concentrate on it, the less you grasp the whole . . . it is such a bewildering chaos of colours that your poor imagination is dazzled as though by continuous fireworks.

(cited in de Botton, 2002: 73)

Camel riding through the desert, Bedouin dances, belly-dancing lessons and laid-back evenings at *shisha* cafés are all rich and colourful components of a fantasized Orient materialized in embodied performances (Figure 6.2, bottom right). Thus, they are illustrative of tourism's entangled and blended geographies of the virtual and material, the fantasized and 'the real', the embodied and the poetic. Thus, the

Figure 6.4 Provisional set-pieces.

'fantastic realism' (Bærenholdt and Haldrup, 2004) of the Orient involves scenes, set-pieces and accessories that afford 'the Orient' to be inhabited by immersion and incorporation.

Second, such playful performances of 'the Orient' do not need 'auratic' qualities to succeed. Bedouin desert camps may be in concrete and iron as well as crowded together (for logistic purposes) (Figure 6.4, top), but their settings, backcloths and accessories afford the play of the 'Oriental night' to be enacted, not as a 'staged authenticity' (MacCannell, 1976/1999) but as an 'authentic stage play' in which tourists can take part corporeally, by engaging in the dramaturgical environment afforded by the set-pieces and accessories of the play. This is further exemplified by the provisional character of many of the set-pieces of 'light Orientalism', in bars, restaurants, hotel lobbies and streets (see Figure 6.4, middle and bottom). In that sense, notions of the 'Arabian Las Vegas' or 'Disneyland Egypt' may be read not only as derogatory nicknames for a faked Orient but as metaphors of performance of play and make-believe.

Touching the stones

Above, we discussed the 'light Orientalism' that decorates enclavic tourist spaces in Turkey, and particularly Egypt. When package tourists depart from these spaces, their destination is often a famous site. Both Turkey and Egypt abound with globally recognized natural and cultural sites that are packaged into the sun, sea and sand holiday. More than any other site in the world, the pyramids at Giza, on the outskirts of Cairo, Egypt, epitomizes this. To MacCannell, the pyramids exemplify how the work of others is transformed into a fetish for tourists by being detached from its social and cultural meanings and reduced to 'monumental representations' that can be endlessly circulated as souvenirs (MacCannell, 1979/1999: 6, 149; see also Chapters 5 and 8). Despite this trivialization, tourists continually embark on long, often troublesome and discomforting, journeys to the Giza pyramids (eight hours' bus drive from Sharm, much longer from Hurghada).

In Chapter 4, we discussed how tourist places such as the pyramids at Giza, Cairo, could be understood as 'floating', mobile places. To emerge as an iconic place, it has to be framed and prefigured through a multiplicity of mobilities, networks and performances in order to translate these 'piles of stone' on the outskirts of Cairo into a tangible material *and* symbolic place. Hence, the transformation of the pyramids from simply being big 'piles of stone' to a globally recognized heritage and tourist sight depends on contingent networks of technologies (buses, the internet), representations (pictures, films, home pages), materialities (stones, souvenirs, fences, parking lots), bodies (camels, tourists, traders and drivers), and stories (by experts and tourists, high culture or popular culture). At the plateau, there are ongoing tensions about the 'framing' and choreographing of tourist performances. In essence, these tensions revolve around whether the Giza pyramids are (should be) a place of sight-seeing or site-sensing. Site management has for years attempted to 'musealize' the plateau for conservational reasons and to preserve (or recreate) the plateau as a *visual* sight, providing parking lots, viewing stations and fences around the pyramids (see Hawass, 1998). Yet visitors have long engaged in more *multisensual* encounters with the pyramids by, for instance, climbing them and riding camels around them.

We joined the company of a family on a two-day sight-seeing trip organized by a tour company to the pyramids and Cairo. This trip was intended as the peak of their winter holiday in a fashionable hotel in Sharm. The family consists of two children (10 and 12 years old) and their mother. The father had not been able to get time off from work, but the mother travelled with the children to Egypt anyway, as it would be 'a pity for them if they [the children] would not be able to travel in the school holiday because of him'. The sight-seeing trip is planned to include, on the first day, all major parts of the Memphis World Heritage Site and, on the second, the Egyptian Museum, the historic quarters of Cairo (another World Heritage Site) and the Citadel. Hence, the trip endorses a strictly regulated choreography of bus driving, stops, guiding, sight-seeing and waiting time. Leaving Sharm el Sheikh in the middle of the night, the bus arrives at the first stop in Memphis early in the morning. After a short stop, we proceed to Saqqara (with the famous step pyramid), and after a lunch break the sublime climax of this trip approaches: the pyramids of Giza:

> [T]he first thing that happens as people leave the bus is to catch their camera; as a reflex response to the view of Khufu's pyramid. Some do not even go out of the bus before they start shooting with their cameras and mobiles . . . They look like gunslingers in a city battle, grabbing their pistols as they confront their target. Unsatisfied with the result, people move across the parking lot and try to get a better 'shot'. I tried to follow our family (discreetly) . . . tongues out of the mouth in pure concentration . . . all three of them shooting . . . before I approach them to compare our results: 'Did you get a good photo?' 'No not really'. Disappointed, we join the guide and the rest of the group.
>
> (Michael's field diary, 19 February 2008)

In Chapter 7, we examine in more detail the performances of digital photography at famous attractions. What interests us here is how photographing becomes a way of relating to the materiality of the site. The Giza pyramids seem to possess a power that pushes the trigger automatically. Arriving at the plateau (Figure 6.5, upper left corner) people immediately start 'shooting', not even attempting to get a good view. When tourists were mostly equipped with analogue cameras, guides would instruct them not to waste their film here. It is simply not possible to take a 'good' picture of the Khufu pyramid from this angle; it will be a picture of nothing but of a 'pile of stones' (Figure 6.5, upper right corner). The iconic images of the three pyramids (see Chapter 5) are produced from the viewing station 'behind' the pyramids (opposite the parking lot). So why do people literally shoot as an affective response to facing these 'piles of stone'? The mother from the family we 'shadowed' may provide an answer:

Figure 6.5 Sensing the pyramids.

We came here for the sun and the sea', the mother explains, 'but this trip was a 'must'. The pyramids!!! Fantastic. Fantastic!!! I mean, you have seen them over and over again, everywhere, on pictures, television, everywhere, and then being there, *touching* the stones, I think, that was why we took so many pictures, and you would not take a stone, would you? That would not be OK. I have never read a single line about them, but NOW when I get home, I will, it's amazing what they have built . . .

(Michael's field diary, 20 February 2008)

It is more the production of a tangible and material 'token' that fuels her photo-graphing than it is an ambition of imitating or reproducing images seen prior to the visit. Considering the (impossible) option of bringing home (stealing) a stone, photographing becomes the way in which the stones can be touched and brought home (see also Chapter 8). Thus, her photographing may be read as a way of establishing the material reality of the Giza plateau by producing an object that *mediates* between the pyramids and their everyday life (including their absent father who was an ever-present character in their discussions of photographs and experiences). This also resonates with common practices of resting on and touch-ing the stones, even 'climbing the pyramids' (strictly forbidden) (see Figure 6.5, lower left and right corner; see also Mitchell, 1988: 25). Such activities contain their own joy of sensing, wrestling, handling, crawling and traversing the stones, engaging in a physical fashion with material objects.

Other tourists

We now move our focus from questions of how people relate to the places they visit to the relations *between* tourists. Issues of sociality are crucial to understanding why and how tourist performances take place (as argued in Chapter 2). The significance of sociality and sociability surfaced on many other occasions and in different forms in the formal interviews, in our informal conversations with tourists, and when observing tourists interacting with each other. In order to bring out the issue of 'sociality', we will mainly draw on interviews made after their return and, to a lesser extent, on field observations.

First, whether we speak of 'package tourists' or more 'independent tourists', our interviews and observations indicate that the desired attraction of many tourists' holiday is the company of their travel companions, their family or partner, as much as, and in some instances more than, the place and its attraction as such. We may say that they desire intimate, sustained and slow-paced dwelling with their significant other(s). It is actually by leaving behind the time-consuming and dividing (e.g. work, school, leisure activities) routines of the everyday and travelling to a new place where they do not have such obligations that partners and families have time for each other, to relax and have shared experiences. In Istanbul, Jonas encountered a young woman travelling 'independently' with her partner across Europe. When we later interviewed her at home, she explained:

> It was just basically a question about being together, because our lives had been very busy especially for the last half year. Before we went on this holiday, we lived as partners in each part of the country. So for that particular month, there is no other than us, there are no mobiles, there are no friends, there is no television, there are no [university] books. It is only us, and we shall talk and have a nice time together . . . read books to each other and things like that . . . So we don't need big 'highlights' [attractions] . . . there were days where we just sat and read aloud to each other almost all the day.
> (home ethnography 1)

For many families and couples, the holiday is a very 'private' moment in which one can try to make up for the absences and lack of intimacy experienced at other times of the year. The holiday destination affords a unique possibility to socialize with your intimate others. Whereas swimming pools, for example, for many (and 'critical' tourism theory in particular) conjure up an image of hedonistic, lazy, flirting sunbathing bodies, for many tourists, especially families, they are also places of much carefree and joyful sociality. Family bodies, young and old, fool around in the water, play with balls, fight, race down or play tricks on the waterslides and jump acrobatically into the water or make 'water bombs'. The swimming pool is an emblematic tourism space of such corporeal sociality (see also Chapter 4).

Second, tourists inevitably bump into other tourists, and forge more or less intense and more or less temporary bonds with co-travelling tourists (something we gained from as researchers). In particular, package tours and inclusive hotels afford meetings and bonds between co-travelling strangers. Here, tourists depart together and dwell – e.g. eat, sunbathe, swim and drink – in close proximity to each other. This 'bonding' begins in the airport, where one can observe how tourists discreetly check each other out and make small talk with those next to them in the check-in queue, the boarding lounge, the plane and the bus ride from the airport to the hotel. And this is unlike most public transport behaviour, characterized by Goffman (1963) as 'civil inattentiveness' – in part, because such travellers depart once they arrive; they share no future or faith together. In contrast, package tourists know that they are, for better or worse, destined for the same destination; they maintain a certain polite attentiveness to each other throughout the journey from the airport to the hotel, and this polite attentiveness continues right through the time at the hotel. There is an expectation of mutual attentiveness, unlike the civil attention amongst 'strangers' in the public of the everyday (Goffman, 1963: 84–5). The morning after their arrival, tourists greet each other at morning buffet (not common restaurant behaviour!), and some exchange small pleasantries about yesterday's journey, the hotel and weather, and how they slept and what plans they have for the day. As the week progresses, it is observable how most couples and families have found 'friends' among the other tourists that they routinely talk and socialize with: they greet and chat when they bump into each other; share travel tips; sit next to each other in the restaurant and on the sight-seeing bus; arrange private excursions together; hang out and play together in the pool; share a coffee in the afternoon and drinks in the evening in the bar area, and so on. For those travelling with children, it is often the children that initiate this contact, as children quickly become 'friends' when playing in the pool area, and this means, especially if the children are young, that their parents also (willingly or unwillingly) spend time in the company of other parents. Reflecting on his fieldwork with his partner and two-year-old son, Jonas recalls:

Due mainly to his [my son] play and friendship with children of roughly similar age, we became part of a group of four very different (discriminated according to education, political observation and cultural preferences) Danish families that increasingly, as the week progressed, talked, played, ate and drank in each other's company, when we routinely bumped into each other or made more or less concrete plans to meet or do things together at a certain time. As the week came to an end, we loosely talked about meeting up in Denmark and phone numbers and business cards changed hands. While the group never met up after holiday (to our knowledge), strong, meaningful and memorable bonds were forged, and it required a ritualistic farewell that extended the bond into the future – at least at the very moment when the exchanges were performed.

(recalled by Jonas, June 2008)

The point here is that these four very different families (although, crucially, with similarly aged children) in part spent time together because they lived in an enclosed tourism enclave where they could not avoid bumping into each other and being proximate. We may say that package hotels and package tours afford such holiday friendships, being places 'packed' with sociality and proximity. Whereas most package tourists appear to enjoy this sociality, some also stress its drawbacks: it can be intimidating and difficult to escape and, not least, disrupt one's *private* family project. For instance, reflecting upon one particular situation, a single mother travelling with her daughter puts it this way:

> First it was okay; we talked for an hour or so, and I told him [similarly aged man] what we had been up to and what we had experienced . . . And when he was a bit like: 'Why don't you come down and talk tonight?' And when I felt: '. . . you spoil my family project, go away'. His friendship was too intrusive.
> (home ethnography 11)

The social game of being on a package tour is also one of tactfully avoiding and declining unwanted travel company. And the more one desires calmness, privacy and a family-oriented holiday, the greater the significance of succeeding in that game. Playing with words, we may say that it is difficult to be exclusive on an all-inclusive tour. It is not only package tourists who make friends while holiday-ing, nor do all tourism friendships turn to dust. Some of the more 'independent' tourists also stress how they occasionally meet holiday friends on the move and maintain some such friendships, for shorter or longer periods, through intermittent phone calls and emails (the latter often with holiday photos attached). Not all holiday friendships are short-lived, evaporating on return home.

Third, many tourists return to the same place year after year. Some of these buy holiday flats. Alanya, Turkey, is a particularly popular place for such second homes among Danish visitors, but Egypt is also home to the holiday flats of many north European visitors. Others simply return to the same place year after year. Once the place's novelty has worn off and the sights have been seen, the holiday experience often becomes a mundane, slow-moving one of bumping into and catching up with fellow repeat tourists and/or holiday flat owners. As one of the couples we interviewed recalled:

> It was exciting when you first came down there and one was eager to go out and try mud bathing and all those things that you were supposed to see and try. Turkish baths and all those things not to be missed. But when you have been down there many times, it is no longer what you search for; you are more interested in coming down and relax. And walk and talk with those people that also are down there. Basically, just having a relaxing, good time and not doing anything . . .
>
> We can never really remember if it is ten or eleven or nine times . . . the last couple of years it has been the same hotel . . . Here we meet that couple from Zealand [Denmark]: that couple with the gold chains. Those we sat next

to at the end. They are here every year. They are not among those that we email with. And then there is this couple whose daughter is married to a Turk. They are usually also down there, but not this year. Then we meet a couple. Was it last year we met them? We had seen them before; we remembered that a year before we saw them at a Turkish Night at a restaurant where we had something to drink.

(home ethnography 15)

Social bonds with other tourists are important in tying together the everyday lives and 'homes' of people with the spaces in which tourism is performed. Other tourists play a significant role as intimate co-travellers, peers to socialize with (occasionally while on holiday or over a sustained period of time), and locals not only take on the roles of 'exotic' characters blending into the tourist scenery but also occasionally become important brokers, gatekeepers and even friends for the visiting tourists. Thus, the sociality of tourist performances contributes in significant ways to transforming the holiday destination into a temporary home.

Home-making

Whether we turn to tourism brochures, articles in magazines and newspapers or academic books, tourism is conventionally portrayed as an experience through which we temporarily *disconnect* from home; indeed, tourism is often explained as an escape from everyday life. And, yet, we have argued that a tourism lens attuned to the everyday communication flows and mobility must also explore tourism performances of what we in Chapter 3 termed home-making. To recall, one aspect of home-making refers to how places afford a homely feeling, and tourists in turn make themselves at home in a foreign place.

One way to make oneself at home in a foreign place is to pack 'homely' objects (such as foodstuff and drinks, communication technologies, games and books) into one's luggage. Our observations and interviews show that almost all tourists carry their mobiles with them on their holiday despite the high roaming charges abroad (see 'Connecting back home' below). Whereas many couples or families often share one or two cameras between them, they bring and use their own individual mobile that does not change hands much. The fashionable iPods are also brought along and much used when people try to kill or speed up travel time and slow down time when they relax at the pool or beach. So, the imaginative travel that music affords is both used to escape and immerse oneself in a given state or place.

To cite Oscar Wilde, the champion of hedonism: 'I no longer want to worship anything but the sun. Have you noticed that the sun detests thought?' (cited in Littlewood, 2001: 190). The stereotypical image of the sunbathing tourist is that it is a state of mind that epitomizes a fleeing from any intellectual activity or sensibility. This is nicely encapsulated by the lyrics of Morrissey's song 'The Lazy Sunbathers' (from the album *Vauxhall and I*), the words of which go:

A world war
was announced
days ago
but they didn't know
the lazy sunbathers
the lazy sunbathers

Nothing
appears
to be
between the ears of
the lazy sunbathers
too jaded
to question stagnation.

(Morrissey, 1994)

Although the sunbathing tourist may be an apt metaphor for political and intellectual laziness and ignorance, these 'lazy sunbathers' with apparently 'nothing between the ears' are not 'too jaded' to consume numerous fiction books. Pools and beaches are not only places of 'Dionysian' tourism that celebrates sensuality, abandoning and intoxication (Rojek, 1995: 80). Our observations, diaries and interviews reveal that many tourists take novels with them on a holiday. A major attraction of a holiday in the sun is that it affords time and space – not least in a deckchair in the shade provided by parasols at the pool or beach – to read books and thus enjoy imaginative travel, that is travelling imaginatively through texts and images. The holiday is probably the time of the year when most works of fiction are read. Interestingly, the many 'reading tourists' we interviewed and observed reading at beaches and pools seldom read guidebooks, and their novels rarely related in any particular way to the country or culture they were visiting (Figure 6.6). Novels are used not as a door into a more poetic connection and immersion with their new home, but rather to travel elsewhere, to other fictional worlds and their places and times, through what we choose to call – playing on the notion of armchair travel – 'deckchair travel'. Tourists do 'deckchair reading' because they love to read and often have too little time to do it at home.

Everyday lives are often structured around and routinized through the television set and certain national programmes at a specific hour. Such television routines are likely to be disrupted when we go abroad. Many charter tourism hotels do not have a television set in the first place, and those that have normally show international channels. For some tourists, this is a little annoying, and they solve the problem by bringing their own portable DVDs and films with them. As one single woman travelling with a friend says:

I have one of those small portable DVDs. I actually bought it because I, we, were going on a holiday . . . I bought it because we have so much time in the evening . . . so we can sit there and watch some films . . . we have learnt that

Figure 6.6 Homely objects.

from experience . . . we once stayed at a nice hotel but it had no film channels or anything like that. We don't go on a holiday to get drunk and stuff like that. The DVD is a pleasant way to pass the evenings.

(home ethnography 11)

A father travelling with his wife admitted that his two children were allowed to watch Danish cartoons at the same time in the afternoon as their usual children's programmes back home; the children need to disconnect a little so they connect the DVD player:

We brought our DVD player with us. Yes, so that when we came up to the room late afternoon, the kids could relax a little bit . . . they need to get disconnected once in a while.

(home ethnography 10)

Another persistent stereotypical image of organized mass tourism is that of the tourist who cannot leave home without bringing along supposedly indispensable foodstuffs, fearing or knowing from past experience that they cannot be purchased at the destination and that they are crucial components in orchestrating the good holiday and a home away from home. Yet our interviews and observations showed that few tourists today – even those in holiday flats – take much food with them. Exceptions to the rule, in the case of Danish tourists, are the ubiquitous dark rye bread, followed by filter coffee (of the preferred brand) and for some, especially older tourists, the Danish bitter Gammel Dansk (Old Danish), all essential ingredients in many Danes' daily or weekend morning and lunch rituals.

How can we explain this? A 'cosmopolitan' explanation would be that, as the tourists of today are experienced travellers (compared with the early era of charter tourism) and ethnic food is served and sold in even small provisional towns throughout Western Europe, many are likely to have developed a certain cosmopolitan sensibility towards and curiosity for 'local' or 'exotic' dishes. Moreover, most Western tourists know that, even in relatively exotic places such Turkey and Egypt, Western food is always within reach, at least if one does not venture too far off the beaten track. This ties into the 'McDonaldization thesis' (Ritzer, 1997; Ritzer and Liskar, 1997), according to which supermarkets sell Western/global food while hotels and tourist restaurants serve Western-style foodstuff, Western/global food brands, Westernized Oriental food as well as dishes from tourists' homelands. Thus, it is easy for Western tourists to avoid the tastes of the Orient and to continue consuming their normal food brands without transporting them themselves (see Figure 6.7).

During both trips to Alanya, Jonas and his family stayed in smaller two- and three-star hotels – made up of one-bedroom flats with a kitchen – on the outskirts of the city. Both hotels are surrounded by other hotels and tourist flats but also by local shops, restaurants and residences. They have a distinct Scandinavian feeling: all the guests are Scandinavian, mostly Danish; the Turkish staff greet and charm with Scandinavian (Danish and Swedish being very similar languages) words; Danish and Swedish flags are displayed; Carlsberg is the favourite beer; many, if not all, the information signs and menus are in Danish; there is a bookshelf with Danish and Swedish weekly gossip magazines and books; Scandinavian tabloid papers are sold; Danish and Swedish religious festivals are celebrated and global fast food dishes are served alongside slightly Westernized Turkish dishes to please Scandinavian tastebuds and alleviate the fear of strange and unknown food; menus include photos of the dishes, and at one of the hotels the menu describes in detail the ingredients (including the exact weight of each) that go into each dish.

Such visual and detailed menus have traces of Ritzer's McDonaldization process and they 'empower' tourists. One aspect of this is the replacement of human by non-human technologies: this is what happens when the images do the talking for waiters and tourists with inadequate language skills. Moreover, such images also ensure what Ritzer calls 'calculability' and 'predictability', as tourists know precisely what they pay for and can expect; they are a guarantee against 'exotic' surprises and 'cheating' behaviour.

The city of Alanya has an abundance of 'tourist restaurants'. Serving a mixture of global and Turkish dishes, and readily identifiable by insistent Turkish waiters on the street, they are characterized by bright colours, multilingual menus containing large images, national flags on the tables, international football t-shirts on the walls, the sound of international pop tunes, large outdoor serving areas and large television screens showing international football matches. They are mostly located on the main streets or in tourist areas and filled with tourists but no locals. In contrast, restaurants frequented by locals, mostly indoors and bathed in neon light, are to be found largely on the side streets; menus are seldom visible, only Turkish food is served, and tourists are extremely rare. The restaurant scene in Alanya is visibly divided and very few tourists eat where the locals go; dining tourists rub shoulders with like-minded tourists in resturant enclaves catering for both international and, especially, national home-coming.

It is striking how 'banal nationalism' permeates many of the restaurants in Alanya. National flags are central components in performances of 'banal nationalism', and they are omnipresent on the Alanyaian restaurant scene; in Figure 6.7 we can see how restaurants called Sunset Copenhagen, Scandinavia and The Viking ('Vikingen') nonetheless have 10 European flags waving next to their names. Tourists are continually reminded of their citizenship and nationality, e.g. by being asked where they are from by waiters who try to lure them into their restaurant. Once the tourist reveals his or her nationality (say Danish), waiters begin to charm in Danish, pointing out that the place is popular among Danes, that cold Carlsberg beers are served or that a forthcoming Danish football match is being shown, or pointing to a Danish football shirt on the wall. And, once you are inside the restaurant, *your* national flag is placed at your table. Football jerseys, whether of the national team or national clubs, decorate the walls of many bars and restaurants, and they too function as markers of national identity. For instance, the restaurant Sunset Copenhagen has the jerseys of rival Danish football clubs on display while Oscar's Scandinavian Restaurant, on the other side of the street, is ornamented with football jerseys of Norwegian teams. Both restaurants advertise on the street that they show Danish and Norwegian matches and they have – like so many other restaurants and bars in Alanya – centrally placed large television screens where they show 'not-to-be-missed' football matches and other sports events announced on the street or just as wallpaper. Thanks to global satellite television, Danes can still follow their local football team abroad.

One interesting finding is that such 'Scandinavian' restaurants tend to be owned by and employ Danes of Turkish origin or Turks living in Denmark outside the summer season. It is in those restaurants with 'Danish Turkish' staff where Danish-ness is staged and performed most.

The streets in Sharm el Sheikh, with its Oriental imagery, are visibly different to the 'nationalized' spaces of Alanya. Yet, here also, restaurants afford the
national and international inclinations that visitors carry with them to be enacted
(Figure 6.7). Although many tourists actually do immerse themselves in Oriental
'otherness' (see 'Oriental nights' above), the encounter with strange tastes,
smells, and so on often seems to be more of a 'tourists' trial' than an expression
of cosmopolitan curiosity. In Sharm el Sheikh, Michael discussed this with the
'shadowed family' who once tried to eat at a 'typical' Egyptian restaurant:

> On the way back from Cairo, I asked 'our family' if they had tried any of the
> Arabic restaurants. Yes, they had. Thursday they ate at the restaurant above
> Hard Rock Café – the one that the guides point out as 'truly Egyptian'. 'It

Figure 6.7 Home-making.

is a thing that you just have to do, and we decided beforehand to do this only once'. The children did not really like it but 'now they can tell it back home . . . that they've done it'. Anyway, it is more convenient to eat at the hotel with two children.

(Michael's field diary, 20 February 2008)

Mooring

Many tourists not only bring with them (or purchase) homely objects and goods to create a homely ambience. They also make more enduring connections with their holiday destination by 'mooring' their everyday life more permanently to it. Many tourists repeatedly return to the same destination, and even the same hotel, year after year. Such repeated tourism can be seen as another strategy of making connections in tourism, to make a foreign place feel like home, of belonging, as is the case with many summer cottage owners (as discussed in Chapter 2). Some tourists are regulars at the same destination and hotel for several years, and, slowly, the place becomes 'homely' for them. By returning to the same destination again and again, people get to know the place; they become familiar faces to the hotel staff and regulars at their favourite restaurants, and know the shops with good bargains and high-quality products, the good spots at the beaches and where to buy their daily necessities. They hardly need any time to find their feet; they head directly into the sun and bump into known faces among the staff or other regular guests. One couple interviewed in Denmark explained that not only do they bump into the same tourists at their favourite resort (Alanya) by chance, they also coordinate their holiday meetings in advance:

> I emailed Henning and Diana and informed that we are about to travel down there. Then Dianna replied by saying that Ingeborg and Jacob are down there at the moment. And they too also always live at the same hotel. We could find them down at the beach because they are always at the same spot.
>
> (home ethnography 15)

Thus, the routinized encounters and being together with familiar faces, bodies and objects help to create a 'homely' ambience in foreign tourist places. Far from being distanced and alienated, such tourist places enable friendship with like-minded strangers and familiarity with places. Over time, such places may soften their symbolic splendour and exoticism. They become 'fields of care' (Tuan, 1996): homely places in which memories and social relations build up over a sustained period of time, a feeling of belonging and care.

There is a strong tradition in Denmark of families owing a second home by the sea, but in recent years there has also been a trend to buy second homes abroad, in places with sunnier climates and were property is cheaper than in Demark, where summer cottages in many places are expensive. Spain used to be the preferred destination, but many people are now buying property in countries such as Bulgaria and Turkey, which have become popular tourist destinations within the last decade. Alanya is one such place, where not only hotels, but increasingly also holiday flats, are erected very rapidly. Outside the city centre, numerous such holiday flat enclaves are already 'inhabited' or are under construction, waiting impatiently for foreign buyers, mostly middle-aged and elderly couples with inherited capital or equity in their house (although this has recently fallen dramatically because of the global financial crisis!). And the town is full of real estate agents catering more or less exclusively for north European tourists; the particulars are normally in English, Dutch or German, and one can usually speak to an estate agent of one's nationality. Despite fighting for the same customers, not all estate agents and tourist operators appear to be in competition. In one of the hotels owned by the Danish tour operator Tyrkiet Eksperten (The Turkey Expert), where Jonas stayed during our research, introduction meetings and free showings for those interested in buying property are held three times a week. Also, since the tour operators generally do cheaper (but less flexible) deals than the airlines, 'flat owners' often fly with the former. The significance of 'connections to home' and what we might term Danish-ness for such flat owners is also evident from the fact that in Alanya alone there are two Danish societies (e.g. http://www.dkalanya.com/).

Tourists also form friendships with 'locals'. Many repeat tourists speak of their friendly relationships with staff at their 'regular' restaurant or shop, where the staff know their face, name, favourite dish and preferred drink, and always greet, treat and talk with them as friends, give them a discount and drive them home to the hotel when the night comes to an end as well as arranging private excursions during the day, in which they also participate. We spoke with several repeat visitors who, because they have developed such bonds and attachment to the owner, staff and place itself, frequent only a few restaurants and bars when they routinely go out for dinner and drinks in the evening. Such places afford what we might term 'friendship hospitality', in which the traditional distinction between friends and guests is blurred. Thus, one of our interviewees, who had during the last couple of years travelled regularly to Egypt (with her two children) to oversee the construction of the two holiday flats they had bought in a resort area, told us that when visiting Cairo she would normally live in the apartment owned by 'friends', an Egyptian family, with whom they had also participated in family gatherings:

> Mohammad and his wife and children they live there, but they move out of the flat when we arrive. The first time we were there they DID set a price, for example, so much for a trip to the pyramids, and things like that. The second time, they would not set a price, so I had to come up with something, but they would only reluctantly take the money. It is something like: It has to be very discrete and not too obviously. I have a fixed amount I give them each day, just to help them get along; you want to be able to help them a little . . . And I have brought gifts for his children and the wives [of him and his brother].
>
> (home ethnography 18)

This blurring between being friend and paying guest also showed up in tourists' accounts of their friendships with the local population. Frequently, our interviewees would downplay the ambivalence of such relations. In some cases, they would maintain contact between holidays by asking their friends for favours such as to make hotel bookings or provide advice concerning travel itineraries, times, and so on. The woman interviewed above maintains regular contact with her friends in Hurghada and Sharm el Sheikh through text messages and phone calls:

> I have four or five people with whom I have regular contact down there. I've got their mobile numbers and things like that . . . I don't know how often, every fortnight or something like that . . . perhaps every three weeks, but it is more like, when you are watching something . . . for example these bombings in Dahab, right, then I phone them . . . Yeah? In the beginning, when something happened I phoned them: if they were OK?, what they thought about it?, how they managed it? Things like that . . . The first time after we had [been there] I called him because, he is travelling a bit around, and he is often in Dahab, so I called, but I guess things are a bit more tense on the Sinai Peninsula, and they are after all living in Cairo so . . . Well they always want to explain things like that away. It's like the first time we were in Cairo. There had just been a bomb, and they . . . well, you know, in the end they will always tell you it is the Israelis. But then you always get another angle on it from them.
>
> (home ethnography 18)

Buying a house and making friends (or family) with people living at the places tourists visited shows how tourism is not detached from other aspects of everyday life. Such ways of 'mooring' (part of) one's life to the place of holidaying erodes the distinctions of home and away and affords a 'multi-locatedness' in which everyday concerns and worries 'haunt' the spaces of tourism, even while being physically absent.

Connecting home

In Chapter 3, we argued that another component of home-making is the use of mobile communication and 'internet moorings' to connect with friends and family members back home. In that chapter, we also discussed research showing that many tourists travelling for a considerable time perform 'connecting home' regularly to alleviate homesickness and stay in contact. We also discussed how global communication moorings such as internet cafés proliferate in tourist destinations, and that most hotels have an internet corner these days. 'Connecting home' is also common among tourists travelling for shorter periods of time. At one of the hotels where Jonas stayed in Alanya, the 'internet corner' is a semi-open space – consisting of six computers – in the lobby area and overlooking the restaurant. The internet corner was particularly busy in the morning as many guests read their local paper electronically, checked their email account or their social networking site such as Facebook or played games. For many tourists, it is a daily ritual to spend a little time on their own getting updated on news, incoming mails and writing email postcards to friends and family members as captured on the photos from Alanya (Figure 6.8).

Overall, our ethnographies and interviews reveal ambivalent attitudes towards mobiles and emailing when holidaying. On the one hand, all the tourists we talked with bring their mobile along but they develop ways of managing their availability and connectivity to home in order to avoid making expensive calls and receive intrusive, disturbing calls. Few of the interviewees state that they continue their normal everyday mobile routine abroad. But there are exceptions:

John:	We use it [the mobile] as we do at home. That's a part of the cost of having one.
Sara:	Around 500 kroner [£50] [on each journey]. Both of us have elderly parents who are ill.
Interviewer:	So you do both texts and calls?
John:	We also call home. My father is turning 80 this March. So to send him a text is a little . . . We have always said that if you have a mobile, you are also allowed to use it; especially when you are far away from home.

(home ethnography 15)

And a woman in her early 20s reflects upon how she used her mobile while holidaying in Egypt:

Yes my mother was dead scared that I was going to be bombed and that she would hear about it in the news, so I texted her: 'I'm down here now and I'm still alive' . . . I texted a lot. My telephone bill came to 1200 kr. [£120] or something like that . . . It costs, it is so crazy, 8–9 kr. [80–90 pence] to send a text.

(home ethnography 11)

Although these interviewees are untypical in maintaining much of their everyday mobile performance, they are typical in the sense that one major reason that tourists bring their mobiles with them is to meet obligations to significant others at home. In Chapter 2, we discussed how much travel demand stems from a desire to connect with significant others and fulfil obligations that cannot be satisfactorily fulfilled through communication. Along similar lines, our observations and interviews show that much of tourists' communicative home-making is tied up with meeting obligations to friends and especially family members back home. This is either because significant others explicitly ask for frequent contact or because it is in the air that such contact is expected, even though it is seldom expressed. It was striking how many tourists text as the plane is about to depart and especially once it has landed. And indeed many of our interviewees report that they send texts shortly after landing to reassure relatives that they have arrived safely. Despite

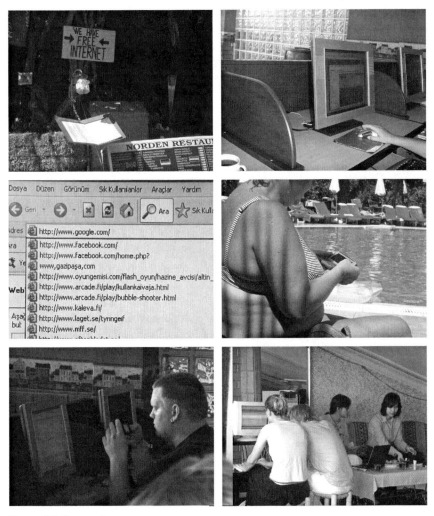

Figure 6.8 Connections.

being a relatively experienced traveller, this woman in her early 60s travelling with her husband explains:

> We send text messages to our daughter . . . She is very much like: 'are you guys alright?'. So we send one from the airport because she is, oh, so caring and would like to know that we have arrived safely . . . she is very concerned about stuff like that.

> (home ethnography 8)

This young couple in their early 20s explain how they call and text shortly after landing to calm down their parents:

Morten: So we called immediately. Both of us. We actually did it in the airport.

Camilla: Yes, because my mother is terrified of flying, so she thinks that everyone is crashing down. So I had to call as soon as we landed. It was not something that she had told me to do, but I knew it was something I had to do. It is just something that I have to do.

(home ethnography 12)

This young woman's way of expressing this obligation is revealing, because it highlights how obligation is not just an external force but also something that individuals internalize. This also mean that individuals will not always recognize 'this is something I have to do' as an obligation, because there has been no explicit verbal instruction.

A long-standing 'obligation' of being a tourist is to send a postcard, and some interviewees highlight how they use their mobile to send 'postcards' by SMS (text) and, to a lesser extent, MMS (photo) instead of, or in addition to, the traditional picture postcard. This is an economy not only of obligations but also of fun and teasing:

We sometimes send messages for fun. To workmates and so on. For instance, we have this workmate: a young man who is 19 years old. And we did it to tease him. He had had his holiday and now it was our turn. And our time to banter: we send him a photo [by MMS message] where we were lying at the pool.

(home ethnography 15)

Indeed, the swimming pool was one of the places where we observed mobile phones and tourists 'texting'. However, making and responding to phone calls was rare; tourists are more likely to text than to talk on the phone, although, even then, they text much less often than they would do at home.

There are various reasons why people call and text significantly less often on holidays than they do at home. The obvious one is financial. High international roaming charges make the use of mobile phones a costly affair. In addition, while travelling, there is less need for the kind of 'micro-coordination' that much everyday mobile phone use is about (Larsen *et al.*, 2008). But what is equally important is that almost all the interviewees express the desire and use of strategies to disconnect temporarily. For some, it was a question of filtering out work-related calls. At a post-travel interview, a male manager was asked if he answered calls on his holiday:

yes, but not work related . . . yes, perhaps I had two calls or something like that, but I always record a message saying that 'I'm on vacation' and that they can call the office instead . . . but my phone was switched on most of the time and if I could see it was a number I couldn't recognise or that it was work related, then I simply didn't answer it. But if it was a private call I would answer it.

(home ethnography 10)

Whereas this particular tourist had his phone switched on and within reach more or less all the time, other tourists deliberately chose to switch it on only at particular times and to leave it in the hotel room for most of the time, because they did not want to be interrupted by calls on a holiday. Although most tourists travel with and use their mobiles *occasionally*, the interviewees highlight that connecting home effected through mobiles is not a major part of their holiday. This was a common response:

We have sent a few text messages to friends and family members, the children have as well, but I also think that when you are on a holiday, then you are on holiday. Then you can wait to speak to people back home till when you return home. It is not important for me to have contact with people back home when I'm on holiday.

(home ethnography 18)

Conclusion

In this chapter, we have traced some of the different ways in which tourism and everyday life are entangled in people's performances while away on holiday. As argued in Chapter 1, tourism is performed in networks that stretch across the spaces of the holiday and the spaces of home. By partaking in tourist travels to Turkey and Egypt, we have elucidated how these networks contingently help to establish these places and the performances tying into them as 'touristic'. The mobilities that tie tourist performances and places together are at the same time imaginative and virtual. Thus, tourism takes place in a blended geography in which materiality and virtuality are continuously entangled. We have shown how encounters with the 'exotic' and with spectacular attractions form part of playful performances in which people incorporate 'other' bodies and materialities in their personal life. Such tourist performances take place in a blended geography of virtual and material encounters, technologized and embodied practices and connect spaces of home and tourism. In the next chapter, we will examine in more detail one particular performance in tourism, that of photographing and how this can be read as a performance. After this, we move into the houses and apartments of tourists to scrutinize the afterlife of tourist experiences and how this is incorporated into the spaces of the everyday home.

7 Performing digital photography

Introduction

In the last chapter we followed in the footsteps of tourists, ethnographically exploring various tourist scenes and performances in Turkey and Egypt. In this chapter we make a detailed ethnography of one emblematic tourist performance – photography – at tourist attractions in Turkey. Photographing is an emblematic tourism performance and yet little research has explored why and how tourists do photography. In Bærenholdt *et al.* (2004) we carried out what appears to be the first ethnography of how tourists *do* photography while sight-seeing at an attraction. Yet we did not discuss in any theoretical detail how we could understand photography as performance, and the rich ethnography in that book is to some extent obsolete since it is a study of how tourists photograph with *analogue* cameras. This chapter, then, theorizes photography as performance and provides a detailed, visual ethnography of digital photography performed at the Blue Mosque, Hagia Sophia and Topkapı Castle Museum in Istanbul and around the castle ruin in Alanya, Turkey.

Whereas much (earlier) writing on digital media resembles *technical* determinism (considering only technical affordances) or *cultural* determinism (exploring only discourses or practices), we take the premise that technologies cannot be separated from embodied practices, from doings, but nor can performances be separated from issues of materiality (see Chapter 1; Couldry, 2004; Shove *et al.*, 2007). The word 'technology', as its etymology reveals, means not only an artefact but also knowledge and craft. Taking inspiration from actor network theory and non-representational theory, we see photography as a technological complex with specific affordances *and* a set of embodied social practices or performances. This chapter considers photography's hybridness of the technical *and* social performances of corporeal humans *and* affording non-humans. Technologies' specific affordances shape but do not determine if and how they can be used and made sense of in practice (Norman, 1999). And this is particularly the case with new technologies – such as digital cameras and camera phones – that are still in the process of being tested and appropriated by consumers. They are to some extent 'underdetermined' (Poster, 1999: 16); they have some 'interpretative flexibility' (Pinch and Bijker, 1984). This also means that digital photography can be many different things according to how camera technologies are assembled, made meaningful and performed in specific contexts, by humans and non-humans. Much research on actual practices of experiencing technologies within especially design studies is overly individualistic, thereby neglecting the rich collective, social and interactive nature of technology use, what the design scholars Katja Battarbee and Ilpo Koskinen (2005) call 'co-experience'. In this chapter we pay particular attention to such co-experiences (or what we prefer to call co-performances) of digital tourist photography, and this is the major reason why Erving Goffman is our major theoretical inspiration.

First, drawing upon the discussions of actor network theory, non-representational theory and performance studies in Chapter 1 and especially an extended examination of Erving Goffman's dramaturgical sociology, we discuss how we can theorize photography through a performance lens. Second, it is suggested that a performance approach to media requires ethnographic research attuned to both human and non-human affordances and enactments. The major part of the chapter reports a visual ethnographic study of digital hybrid picturing practices at tourist attractions and beyond in Turkey. The study examines in detail how the affordances of digital cameras are enacted in practice – how tourists produce, consume and distribute digital photographs – and how tourist photography changed in nature with the transition to digital photography. Although digital photography affords new possibilities (because it performs differently from analogue photography), it is less clear if and how these affordances are used in *practice*, which we set out to explore in this chapter.

Theorizing photography

Photographing is absent from most theory and research, which jumps straight from photography to photographs. Such research typically goes directly to the representational worlds of photographs and skips over their production, movement and circulation. The diverse materialities of photography, hybrid practices and flows of photography are rendered invisible (but see R. K. Cohen, 2005; Larsen, 2005; Shove *et al.*, 2007). It is telling that the opening line of *The Photography Reader* is, 'What is a photograph?' (Wells, 2003: 1). This is also the case in books on visual methods. For example, all the chapters in *Visual Methodologies* (G. Rose, 2001) concern already produced and in-place images in order to analyse their *messages* – their content, meanings or codes (see also Van Leeuwen and Jewitt, 2001). As another example, there are many representational – semiotic or content – studies of postcards and brochures, but hardly any of their production, distribution and consumption in practice (but see M. Crang, 1999; Larsen, 2005). Such representational accounts have been successful in analysing photography as texts and scripts, but they have been blind to issues of technology and hybridized performances, which also means that they have neglected much of the significance of *digital* photography.

Embodied photographic (and media) performances are also absent in approaches that highlight either materiality or cultural discourses. The 'technicist' accounts by media theorists such as Marshal McLuhan and Friedrich Kittler argue that media theory ought 'to focus on the material structures of technologies and the changes these introduce into culture, not the ways in which these are used or the content of the messages' (Gane, 2005: 29). This is effectively media studies without humans as media are said to determine our situation. In contrast, what we might call humanistic accounts neglect the performances of media technologies (their affordances) themselves in their insistence upon arguing that the question of technology is not technological at all but rather cultural or discursive. To cite Kevin Robins (1991: 55): 'the question of technology . . . is not at all a technical question'. Here there can be no study of technology, only of how cultures – through human agency – take possession of them.

The problem with these two positions is that they tend to purify technology by privileging either the technical or the social character of technology (see Chapter 4). This separation between technology and culture is questioned by actor network theory and non-representational theory. Both challenge approaches highlighting either society-shapes-technology or technology-shapes-society. Actor network theory and non-representational theory are 'middle-ground' positions that rephrase the debate from terms such as 'determine', 'impact' or 'shape' towards terms such as 'networks', 'relationality', 'hybridity', 'affordance', and so on. The social and technological are seen as mutually determining. Latour would argue that the dual mistake of the 'materialist' and 'humanists' discussed above

> is to start with essences, either those of subjects or those of objects . . . Neither the subject, nor the object, nor their goals are fixed for ever. We have to shift our attention to this unknown X, this hybrid which can truly be said to act.
>
> (Latour, 1993: 6)

Actor network theory and non-representational theory show how technology, culture and society intertwine and interact in all kinds of promiscuous combinations. And the inescapable hybridity of human and non-human worlds is stressed (Thrift, 1996: 24). These 'theories' refuse to equate agency with intentionality and linguistic competences, and see it instead as 'the capacity to act or have effects'. Agency is a relational achievement between humans and non-humans, 'involving the creative presence of organic beings, technological devises and discursive codes, as well as people, in the fabrics of everyday living' (Whatmore, 1999: 26; see also Bingham, 1996: 647).

Actor network theory argues that a particular technology emerges out of the *relations* between social, natural and technical actors (Michael, 2000: 18). Technologies are hybrids. From this position, photography is thus evidently material *and* social, objective *and* subjective, i.e. heterogeneous. It is a complex amalgam of technology, discourse and practice. Photographic agency is a relational effect that first comes into force when a heterogeneous network of humans and non-humans is in place, as Latour (1991) shows in his analysis of Kodak. In other words, 'we need to consider its technological, semiotic and social *hybridness*; the way in which its meanings and powers are the result of a mixture of forces and not a singular, essential and inherent quality' (Lister, 1995: 11). One way to illustrate what it means to speak of photography as a hybrid is through Latour's hybrid of the 'citizen–gun' (Michael, 2000: 26). For Latour, it is neither the person nor the gun that kills, but the 'citizen–gun', the hybrid. Similarly, neither photography technologies (cameras, mobile phones, computers, printers, and so on) nor the photographer makes pictures: it is the hybrid of what we might term the 'networked camera–tourist'. Photographs are man-made *and* machine-made. So we need to study the affordances that new photography networks afford but also how these specific affordances are performed in practice by 'networked camera–tourists'. It is this complex position that this chapter aims at.

However, one shortcoming with actor network theory studies is that they give scant attention to concrete, lived and embodied practices of hybrids. In his other-wise sympathetic review of actor network theory, Dant (2005: 81) highlights that in concrete actor network theory studies:

> it is noticeable that there are very few accounts of the perceptual or tactile interaction between humans and objects in the network, few detailed field observations, photographs or use of video to study the process of the network that would allow the material objects to have a presence in the accounts.

Another shortcoming is that they tend to grasp technologies as exclusively per-forming practically, and they therefore neglect that most technologies enable a wide range of 'function-expressions' and 'opportunities' for performing 'taste and self-identity' (Michael, 2000: 35–6). And Thrift and Dewsbury (2000: 420) note how: 'one of the most interesting projects currently is to try and make actor network theory perform in such a way as to value the importance of things made in performance and to make performance more central to actor network theory'. In the next section we discuss theories that can help us to make performance more central to our understanding of hybrid photography.

Perspectives on photography as performance

To develop a conceptual framework of photography as a performance we will get inspiration from non-representational geography/media studies, the discipline of performance studies and especially Goffman's dramaturgical sociology.

Let us begin with the first two positions discussed in detail in Chapter 1. To put it briefly, both non-representational theory and performance studies are concerned with bodily doings and technical enactments rather than representations and mean-ings. They challenge the textual dominance in the social sciences and humanities by being concerned with performative presentations rather than merely represen-tation and meaning (Thrift, 1997: 127). They are concerned with 'performances of the now' and with the 'more-than-the-representational' (Lorimer, 2005).

Both a non-representational and a performance studies approach to media studies move the focus from consumption to how ordinary people, as creative, expressive, hybridized beings, use media technologies and produce media products, such as telephone calls, emails, music, film, web pages and photographs (Larsen, 2006). In this sense they have much in common with the media theorist Nick Couldry's (2004) calls for a new paradigm in media studies that is to explore media as open, eventful practices and not as already in place texts or structures of production. This involves a turn of attention from abstract readings of affordances to situated practices of media. As Cooper *et al.* (2002: 286–7) say: 'A technology may have, or be conceptualized as having particular potentials, but the latter's realization – or reconfiguration, or subversion, takes place in and through, and thus depends on actual situated everyday practice'. Or to cite Norman (1999: 41): 'the only way to find out what people do is to go out and watch them: not in the laboratories, not in the usability testing rooms, but in their normal environment'.

The practice of taking photographs is often conceived of as a visual practice that is rapid and does not require much work ('just a press on the button'), but non-representational theories and performance studies approaches can help us to highlight busy, active, playing and active 'bodies of photography'. When we conceive of photography as a performance, it is a process over time. Indeed, our ethnography of analogue tourist photography shows how analogue photographs are rarely the outcome of a quick shutter release. For instance, bodies of photographers that stand erect, kneel, bend sideways, forwards and backwards, lean on ruins, and lie on the ground are constantly visible at tourist attractions. Much tourist photographing is enacted, lengthy embodied visions involving touch, body language, 'face work' and talking (see Bærenholdt *et al.*, 2004; Larsen, 2005).

According to performance studies, all performances contain elements of play and ritual (see Chapter 1). Normally, photographing is seen as a means to an end (photographs), but the play aspect turns things on its head: photography can now also be an end in itself. Without neglecting the value of photographs, the play aspect can help to show how photographing can be a source of pleasure, creativity and sociability (in Simmel's sense; see Chapter 2) in itself, and this may in part explain why photography is an all-consuming performance of tourism. There is a significant *play* element to photography, but this fact is often drowned in writings highlighting the ritualized nature of photography and what it represents. One exception to the rule is Orvar Löfgren. Writing at a time before the invention of digital cameras he states:

> The critique of the urge to document misses an important point. The pleasure may not be in gathering up moments to display next winter but just in creating them: Letting the video roll . . . clicking through a roll of Kodachrome. However much energy goes into the production of these narratives and whatever their fate, producing them was an experience in its own right . . . Here is an arena where nonartists . . . do not hesitate to try their hand at producing, a photo narrative . . . [or] video documentary. Here you may become your own director, scriptwriter or scenographer.
>
> (Löfgren, 1999: 74)

Thus, whereas *representation* used to be a fitting basis for tourist photography theory, the *more-than-representational* now seems a more illuminating and complex concept to work with.

The ritual aspect entails that we also have to investigate the cultural scripts that frame how we can picture and the meanings and roles that photography has and plays for modern societies. In the sociological literature on family photography, it is agued that the ritual role of photography is that of fixing or solidifying the modern family (e.g. Bourdieu, 1990; Holland, 2001). But the literature does not discuss in detail how families in practice present themselves photographically by enacting various ritualized dramaturgical practices. Yet our third source of inspiration, Erving Goffman's dramaturgical metaphors, can be helpful in bringing into the fore how the 'affectionate holiday snap' is enacted and produced.

Erving Goffman has commonalities with performance studies (in fact, Goffman is included in readers and introductions to performance studies; see Bial, 2004; Schechner, 2006). In *The Presentation of the Self in Everyday Life* (1959) Goffman famously outlines a 'dramaturgical' framework to describe everyday face-to-face social encounters and interaction. Using dramaturgical metaphors, he examines how a person in the company of others 'presents himself and his activity to others, the ways in which he guides and controls the impression they form of him, and the kinds of things he may and may not do while sustaining his performance before them' (ibid.: 8). Goffman examines how individuals and groups – consciously and unconsciously – perform for each other and through inter-subjective dramaturgical practices attempt to *give* specific signs about themselves and the social situation that they are part of.

Being-in-the-world and social life are fundamentally performative and, metaphorically speaking, put on stage for an audience. Goffman builds on the hypothesis that, when people meet in public stages, each individual will immediately seek to control the impressions of themselves that the others inevitably will read off, and they do so by presenting him/herself in accordance with approved cultural norms and expectations. Goffman makes a distinction between to 'give' and 'give off' bodily information. To 'give' is intentional communication whereas to 'give off' refers to the fact that we always, willy-nilly, emit signs when we are in the presence of others (Goffman, 1963: 14). Performances strive to 'give' particular signs and avoid the 'giving off' of uncontrollable signs.

So, rather like performance studies (see Chapter 1), Goffman's performing self is simultaneously creative, strategic and calculating and yet embedded within a morally constraining universe of appropriate cultural norms. In particular, Goffman stresses how performances are about not harming others' faces or causing scenes that damage others or the situation itself. Goffman is interested in the sociological self, and he refers to Park, for whom the etymological meaning of person is 'mask' (Goffman, 1969: 30). The self, for Goffman, is social and interchangeable, a dramaturgical *effect* of performances before a 'judging' audience, the latter having the power to accept or reject the performer's self-presentation. To cite Goffman at length:

a correctly staged and performed scene leads the audience to impute a self to a performed character, but this imputation – the self – is a product of the scene that comes of it. The self, then, as a performed character, is not an organic thing that has a specific location, whose fundamental fate it is to be born, to mature, and to die: it is a dramatic effect arising diffusely from a scene that is presented, and the characteristic issue, the crucial concern, is whether it will be credited or discredited.

(ibid.: 244–5)

The self is a product of exchanges. Performances cannot be explained by referring back to people's selves; rather, the self emerges through performances. The self is a dramatic effect, that is it is continuously created in *public* performances. Thus, the self is relational. As Goffman says, the self's proprietor is 'the peg on which something of collaborative manufacture will be for a time' (1959: 245). This also explains why the self for Goffman is both a *performer* and *character*: a performer, because the individual actively and corporeally performs his self for an audience by using 'impression management' (it is a 'doing'); a character, because the self is an effect of socially sanctioned flows of performances.

Goffman illuminates how performances entail audiences (now or later, real or imagined) (ibid.: ch. 2) and are socially organized in 'teams' and spatially on 'front-stages', where 'masks' are deliberately put on for a specific audience, which is in contrast to 'back-stages' where public masks are lifted (ibid.: 114). In this sense, self-presentation is always a social game involving others, so we should speak of *co*-performances rather than individual performances. Goffman makes the case that performances are often about presenting one's true face, but *idealization* is another common dramaturgical practice (Goffman, 1959: 47). Goffman mentions how:

Given the fragility and the required expressive coherence of the reality that is dramatized by a performance, there are usually facts which, if attention is drawn to them during the performance, would discredit, disrupt, or make useless the impression that the performance fosters. These facts may be said to provide 'destructive information'. In other words, a team must be able to keep its secrets and have its secrets kept.

(Goffman, 1969: 141)

Goffman's metaphors of 'impression management', 'give'/'give off' information, 'idealization' and 'front-stages'/'back-stages' are productive when thinking about rituals of tourist family photography. Goffman's point is that we all play roles and have done so since childhood. But when we are faced by the camera lens we become extraordinarily aware of our body and its appearance, and we pose by reflex to 'give' an appropriate façade. As Roland Barthes said in an almost Goffmanian fashion:

I have been photographed and I knew it. Now, once I feel myself observed by the lens, everything changes: I constitute myself in the process of 'posing', I instantly make another body for myself, I transform myself in advance into an image.

(Barthes, 2000: 10)

Poses are integral to photography. It seems to be a 'law' of photography that we pose when 'camera faces' gaze at us. When one is being photographed one cannot avoid 'giving off' information, but through posing one can try convey a specific image for the future. It seems to be the case that most photographic situations are typified by (culturally specific) posing (one example of 'impression management') (see Larsen, 2005, 2009). A Goffmanian interpretation of self-presentation in photography would be that we do pose in a particular fashion not because we have a certain self but because the situation calls for a specific pose: poses do not reflect as much as they bring forth selves.

And this seems to be specifically the case with personal photography, a kind of photography intimately tied up with joyful moments and happy faces. Photography is part of the 'theatre' that people enact to produce their desired and expected self-image and togetherness, wholeness and intimacy with their partner, family and friends (Kuhn, 1995; Hirsch, 1997: 7; L. Smith, 1998: 16; Holland, 2001). Frictions are almost automatically put on hold, and even dead gatherings become full of life when the camera appears. Virtually all picturing amounts to a 'front-stage' of encoded and enacted 'impression management'. As Anette Kuhn (1995) so poignantly exposes in her book *Family Secrets*, family photographs can be full of secrets, secrets that are hidden (more or less convincingly) from the viewer. Even families that show little affection tend to perform affectionate family life for the camera. This also means that photography produces *new* rather than mirror situations, which is in line with Goffman's idea that the self is a dramaturgical effect emerging through the enactment.

The fact that photographing often involves 'teamwork' and audiences also indicates the usefulness of studying it *as* performance. Photographing is typified by complex social relations between photographers and 'posers', present, imagined and future audiences. It is common that 'posers' are instructed by photographers or other members of the team to bring into being certain appropriate fronts (the most common being: 'Smile'!) or break off inappropriate fronts or activities. There can thus be conflicts between the team members about what poses are appropriate, say between an 'instructing' mother and her 'posing' teenager daughters about whether their impression should signify pleasantness or coolness. But in most cases teams co-produce *one* social body that is ceremoniously displayed. Everyone expresses respect for the photographic event by posing in a dignified way; gentle smiles are worn, bodies are straightened, hands are kept at sides. No one pokes fun or monopolizes attention (for ethnographic evidence of this, see Bærenholdt *et al.*, 2004: ch. 6).

Much posing is also performed in teams. Touch – body-to-body – is an essential dramaturgical practice in relation to especially family photography (especially to what we elsewhere have called the 'family gaze'; see Haldrup and Larsen, 2003). When cameras appear, almost as a reflex people assume tender, desexualized postures, holding hands, hugging and embracing. 'Arms around shoulders' is the common way of bonding friends and family members as one social body. Such staged intimacy tends to come to an end when the shooting has finished (it would be rather inappropriate to carry on hugging even a good friend once the photo is taken!). To produce signs of loving and intimate family life, families need to enact it physically, to touch each other. To produce signs of affections they need to be affective. Signs of affections equal affections (signifieds) in family hugs. Image is realness. Body-to-body experiences cause, and indeed are a sign of, unprecedented moments of intimacy and love. 'Touch is above all the most intimate sense, limited by the reach of the body, and it is the most reciprocal of the senses, for to touch is to be touched' (Rodaway, 1994: 41). Touch acts directly upon the body. Therefore, being touched can also be an embarrassing experience, repulsive even. So tourist photography produces unusual moments of intimate co-presence rare outside the limelight of the camera eye.

In this section, by drawing upon non-representational theories, performance studies and especially Goffman's dramaturgical sociology, we have outlined a performance approach to tourist photography, i.e. an approach that explores how competent, active, expressive and physical bodies picture places and each other in teams comprising photographers, posing models and present, displaced or future audiences. Moreover, a performance approach is to conceptualize photographing as a process over time that is spatially organized on front-stages (even when they take place on traditional back-stages, e.g. the summer cottage), involve 'strategic impression management' such as staging and posing and – when picturing with *digital* cameras – communal immediate consumption of photographs (see below). Concrete ongoing photography performative practices, such as instructing and posing, the composing of scenery and use of cameras rather than 'dead' photographies, can be given prominence and approached *as* performance. One advantage of this approach is that it does not a priori reduce the significance of family photography to an ideological reproduction of the loving nuclear family as is widespread in the sociological literature of family photography (e.g. Bourdieu, 1990), but instead explores *how* people, as hybridized beings, do photography and present places and themselves photographically. It thus represents a shift from why to how, from studying functions of photography to actions of photography (that might reproduce discourses of loving family life) and, crucially, such performative actions are both representational (e.g. posing, self-representation and drawing on cultural discourses) and non-representational (they involve interactions, work, sociability), that is all photography performances are always more-than-representational. There is more to photography than photographs.

Digital photography

It is increasingly evident that the materiality and network of personal photography are changing dramatically with the emergence of digital technologies. Cameras and photographs are digitalized and they *converge* with new media technologies such as the internet, emails and mobile phones (Lister, 2007). State-of-the-art digital cameras feature wifi technology for instant sharing at a distance.[1] In 2004, some 68 million digital cameras were sold worldwide.[2] The same year Kodak stopped selling traditional film cameras in North America and Western Europe.[3] It now sells products such as digital cameras, editing software, printers, photographic paper and services such as online printing service and printing kiosks. And digital photography is converging with mobile phone technology. In 2004, 246 million 'camera phones' (mobile phones incorporating digital cameras) were sold worldwide – nearly four times the sales of digital cameras. However, while camera phones lag behind in the ever-increasing megapixel race, many mobiles now come with 3.2-megapixel cameras or better, which is sufficient to produce high-quality (smallish) photographs. And mobile phone commercials (see, for instance, Nokia[4]) increasingly highlight the camera. Whereas 'analogue photography' was directed at a *future* audience, pictures taken by camera phones (and digital cameras with wifi technology) can be seen instantly not only by co-travellers but also by people at a distance using mobiles equipped with MMS. MMS messages travel 'timelessly' and, if the receiver has the camera phone at hand, he or she can see (and reply to) pictures of events unfolding more or less in real time (Gye, 2007; Hjorth, 2007; Villi, 2007). In the UK, where around one in two mobiles are camera phones, '448,962,359 MMS picture messages were sent in 2007, the equivalent of 19 million traditional (24 exposure) rolls of camera film'.[5]

Whereas analogue cameras depend upon high-street developing to make their photographs 'come to life', digital cameras make them by themselves and display them *instantly* on the (variably sized) *screen* on the back of the digital camera or front of the mobile phone (see Figure 7.4, below). Whereas analogue photographs always depict *past* events taking place *elsewhere* – what Barthes (2000) calls 'that has been' – digital cameras' screens can also show ongoing events right here, when the spaces of picturing, posing and consuming converge. Digital photography is typified by 'instantaneous time' (Lash and Urry, 1994) or the 'power of now' (Villi, 2007). They seem designed for a late modern consumer society, a '*now* society' in which pleasures need to be immediate and the deferral of gratification is seldom acceptable (Bauman, 1998).

The screen affords new sociabilities for producing and consuming photographs (see Larsen, 2008). In terms of production, the screen can turn photographing into a social and collaborative event because 'onlookers' can also monitor the screen when picturing takes place and the result is immediately available for inspection and therefore for comments from 'onlookers' (who may turn into 'co-producers') and 'posing actors', who may demand deletion and retake(s).

Another – but related – difference relates to possibilities of deleting. With analogue cameras, every 'click' materializes as a material object (if the film is developed!) but 'images' that do not charm at first glance on the digital screen can be erased and retaken at no extra financial cost. This flexibility affords more casual and experimental ways of photographing. Viewing, deleting and retaking integrate with taking photographs, which make it easier (yet time-consuming because of *re*taking) to produce images that live up to the postcards, ideals of loving family life or desired self-images. The *flexible* digital camera seems like a treat to a consumer society saturated with fashion, lifestyle magazines, commercials, models, celebrities and reality shows in which ordinary people become overnight celebrities. In such a society, what Goffman called 'the presentation of the self' takes on renewed importance, and digital cameras satisfy people's desire to be able to *control* how they are presented photographically (Larsen, 2008).

Photography's networked convergence with mobiles and the internet means that the technical *affordances* of photography expand dramatically: tourists can send 'live postcards' of *happening* experiences rather than memories thereof to absent others (MMS messages) so that experiences can be communicated and consumed at a distance in real time. MMS messages can include text, voice and video clips as well as pictures and can be circulated among users with MMS-capable handsets. Most MMS messages contain photographs, and they are occasionally accompanied by text. In the latter case they resemble postcards, although they travel much faster. Computer-networked photographs can be deleted, improved upon through editing, distributed freely and timelessly as email attachments to friends and family members or exhibited on family home pages, blogs or social networking sites such as MySpace and Facebook or through photo/video sharing services such as Flickr and YouTube, which illustrates how networked digital photography is a crucial component of the so-called 'Web 2.0 revolution' (this is developed in Chapter 8).

All this indicates that digital photography is a complex technological network in the *making* rather than a single fixed technology. The newness of digital photography relates to the digitalization of images, media convergence and new performances of sociality (relating to broader social shifts towards real-time, collaborative, networked sociality at a distance). In other words, the affordances of digital photography potentially make photographic images instantaneous, on the one hand, and mobile, on the other hand.

Visual ethnographies of photographic performances

Our ethnography takes place in the summer of 2006 and 2007 at classic tourist attractions in Turkey: the Blue Mosque, Hagia Sophia and Topkapı Castle Museum, all in Istanbul; at the castle ruin in Alanya; at a charter tourism hotel in Alanya (especially at the internet café and swimming pool); and in a budget hotel in Istanbul (especially around the internet-connected computer centrally placed in the dining room). It is, however, an ethnography not of these places as such, but of how couples, friends and families, as teams, perform digital photography and present themselves photographically *within* these places; our 'findings' can thus be generalized beyond these places.

Ethnography is particularly suitable for conducting studies of tourist photography as performance because one of the central commitments of this method is 'to be in the presence of the people one is studying, not just the texts or objects they produce' (Miller, 1997: 72). In Chapter 3 we discussed how observations of events as they unfold are characteristic of ethnography. Inspired by visual sociology and anthropology, the present study of digital photography is primarily based upon ethnographic performances of observing, filming and photographing 'photographing tourists', as they go about filming naturally and with their own cameras and mobiles. In contrast, much of the research on camera phones undertaken by technology design researchers has been quasi-experimental (e.g. respondents are asked to take and send photographs with a mobile provided by the research team regardless of whether they do so normally) and focusing upon improving the technologies and anticipating future use (e.g. Koskinen *et al.*, 2002; Van House *et al.*, 2004, 2005; Koskinen, 2005). As a result, the empirical part of this chapter is largely structured around our ethnographic photographs. Observation is not passive perception but an active and theory-informed *doing*: it is never from anywhere but always from *somewhere*. We always look for *something*, not anything. To a certain degree at least, we always find what we look for (if we know what to look for). And nothing or anything if we have no guiding theoretical lenses. As Susan Sontag notes with regard to the supposedly 'mirroring' camera: '[n]obody takes the same picture of the same thing', so 'photographs are evidence not only of what's there but of what an individual sees, not just a record but an evaluation of the world' (Sontag, 1978: 88). This fact calls for reflexion. Our specific 'somewhere' is the idea of photography *as* performance. We claim not that photography necessarily *is* a performance, but only that the lens of performance is, hopefully, one of the more illuminating ways to understand hybridized doings of digital photography in practice. The 'something' we look for is thus various performances of digital photography: we look at photography through a performance lens. We take a reflexive approach to visual ethnography that acknowledges the subjective, constructed and partial nature of our observations and ethnographic photographs that we publish and the knowledge produced through them (Harper, 2000, 2003; Ball and Smith, 2001; Pink, 2001; G. Rose, 2001; Wagner, 2002). Although we aim at 'documentary' photography, the films and photographs produced are certainly not passive mirrors of reality. Their slices of 'reality' are something made, shaped by our particular 'way of seeing'. We have chosen what to photograph and which photographs to publish. In this sense, they are evidently only partial and incomplete 'documentation' (Clifford, 1986).

We make use of cameras and photographs because of their power to visually record material culture and the physiognomy of performances and social interaction. In this sense we agree with Ball and Smith (2001: 309) when they argue 'that the camera's value as an ethnographic tool is similar to the audio tape recorder: it provides an accurate trace of events that still leaves an enormous scope for analytical interpretation'. The idea is not so much to document 'this is what is', but to make photographic series illustrating how practices of digital tourist photography can be grasped *as* performance, and how tourists use cameras *as* performative tools. It is our *creative* way of picturing with bites of reality – as Roland Barthes (2000: 76) famously said: 'In a photograph I can never deny that the thing has been'. However, there is no such assurance with digital photography since photographs are malleable and changeable in editing programs (see Lister, 2001).

Like the performances and scenes examined in Chapter 6, we perform our visual ethnography by secretly observing, filming and especially photographing photographers and posers in action (although we film much and living images better capture the liveliness and temporalities of photography performances, this material can unfortunately not be presented in a book like this one). We pretend to participate in the event as photographing tourists (with ordinary pocket-sized digital cameras), but our camera is aiming not at buildings, views or our family members, but at *other* tourists and their cameras. We waited for interesting (e.g. photographing that involved more than a single click on the shutter-button) *and* camera-friendly (both informative and aesthetically pleasing!) 'team perform- ances' passing by at a particular spot or we 'shadowed-in-walking' photographing tourists who are unaware of our 'shadowing': they do *not* pose for our camera, and we never ask them to do so. So it required patience, luck (being at the right place at the right time, having the right light, etc.) and some cheekiness to produce the photographs included in this chapter.

To bring out tourists' own accounts of their photography performances, 23 semistructured (half of them with Danish tourists) qualitative interviews were conducted mainly at the *exit* of the attractions so that people's photo performances were fresh in their mind. They lasted from 5 to 30 minutes. We now report our ethnographic study, beginning with the more technical aspects of the ethnography.

Technical landscapes

Digital cameras

When Larsen (2003, 2005) and Bærenholdt *et al.* (2004) made a visual ethnogra- phy of tourist photography in 2001 and 2002, digital cameras were a curiosity ('a male jewel'); in 2006 and 2007 it is analogue cameras that are a rarity at tourist attractions. Reflecting the dramatic penetration rate of digital cameras within recent years (to the degree that they have become mundane), our observations at the attractions in Turkey clearly show that virtually all Western tourists today use some sort of digital camera. In five years it has become a rare sight to observe photographers placing cameras in front of a eye; the ubiquitous sight is cameras held at half-arm's length from the face and 'photographic seeing' through the screen by both eyes (Figure 7.1).

The interviewees gave various reasons why they now use digital cameras (all of them have travelled with an analogue camera before): photographs are imme- diately available for consumption after the shooting and retrievable throughout the journey since camera phones and especially digital cameras – unlike analogue cameras – can store substantial numbers (depending on the size of the memory card) of photographs that can be viewed anywhere and at any time; unpleasant photographs can be deleted; there is no worrying about the cost of development or running out of film; photographs can be improved through computer editing

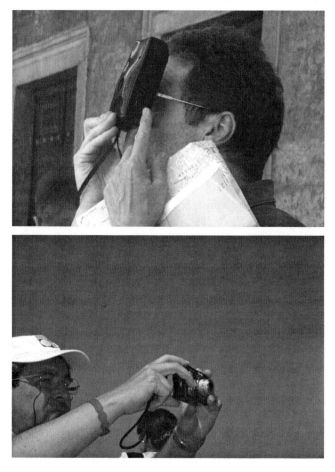

Figure 7.1 Analogue camera and digital camera.

and stored on computers; and digital cameras are cool gadgets and they are fun to picture with. Most tourists have memory cards with a high storage capacity (512 megabytes or 1 gigabyte being the norm), and several travel with more than one card, so that they can take and store (depending on size and quality) between 500 and 1,000 photographs before needing to delete photographs or burn them on to a CD-ROM.

As discussed elsewhere (Larsen, 2008), if tourists connect their camera to the internet, they can send photographs to people back home or upload them to blogs or social networking sites (for instance Facebook) while on holiday. There are internet cafés (or 'corners') in most hotels in Alanya and Istanbul. Our observations suggest that, although tourists use the internet for emailing and to catch up with the news (see Chapter 6), few package tourists send or post their photographs while on holiday. Some we interviewed said that they forgot to bring the necessary cable, but most reported that this was not something that they had considered. But there are exceptions. In the package tour hotel, in Alanya a few teenagers posted images on their social networking sites, and in the independent traveller hotel in Istanbul several 'travellers' (people travelling for a month or more) had brought laptops with a wireless internet connection and shared photographs with people back home. As two of them say:

> We like to share what we are doing with other people . . . every place we go we make a little summary of what we are up to.
>
> (couple from Ireland and New Zealand, late 20s, interviewed in Istanbul)

> We edit in the evening and we have a website, so we put all of our photographs on a website so that people who want to know what we are doing on our trip can see what we are doing . . . post every week or two and we do a little story commentary.
>
> (American couple, early 50s, sailing for several months, interviewed in Alanya)

This suggests that it is primarily those who travel for *long* periods that 'connect photographically' with absent friends and family members back home by blogging or sending photo-messages by email. The reason for this may be that they have more 'free' time on hand and that friends and family members expect them to stay in contact during their trip.

Camera phones

Somewhat surprisingly, given that most mobile phones today have in-built digital cameras, our observations and interviews show that very few *Western* tourists use their mobiles as cameras. Those who do are mainly teenagers and young adults (Figure 7.2).

This finding is even more striking given that, as our interviews reveal, most tourists have their mobile at hand when sight-seeing. Many interviewees complain that mobiles take poor photographs. This is a common remark: 'Yes I have [a camera phone] but I have not really used it – it takes bad photos' (American school teacher, late 20s, interviewed in Istanbul). This is partly because most of our interviewees have relatively 'old' mobiles with cameras with few pixels (e.g. 0.3 megapixels). But there is also a non-technical explanation. Some interviewees do not regard their mobile as a camera and it seems wrong for them to take photographs with it, especially on holiday, when most travel with a proper camera too.

And, indeed, other research shows that mobiles are mainly used for spontaneous everyday photography rather than tourist photography. Research in Japan by Daisuke and Ito shows that:

Figure 7.2 Camera phones.

in comparison to the traditional camera, which gets trotted out for special excursions and events – noteworthy moments bracketed off from the mundane – camera phones capture the more fleeting and unexpected moments of surprise, beauty and adoration in the everyday . . . whereas the capturing of travel photos was down the bottom of the list.

(cited Goggin, 2006: 145)

More generally, in his review of the research on camera phones, the communication scholar Mikko Villo (2007) concludes that most studies highlight that the significance of camera phones is that the everyday is now routinely pictured. It seems that camera phones today primarily capture everyday life rather than tourist life. Only a few tourists send 'live postcards' ('we are *here*' right *now*) by 'timeless' MMS messages (Bell and Lyall, 2005). This is partly because they are expensive and partly because many people do not have the means to receive such messages. This is confirmed by the fact that text or SMS messages far outnumber MMS messages. For example, in January 2008, 42.52 million MMS messages were sent in the UK, compared with 6.3 billion SMS messages.[6] As one middle-aged Danish woman we interviewed recalls: 'I tried to send an MMS message once on a holiday but it didn't make it to the receiver in Denmark. In fact we tried for some 10 seconds to send it. I realized later that every attempt was registered on my telephone bill.' Only one of our interviewees used mainly a mobile phone to take photographs. He prefers his camera phone because:

> I have a digital camera but the camera phone is a bit more handy. It interacts a little better with my computer, so I can transfer photos to my computer . . . With my digital camera I need to transfer photos with a USB cable, as it does not have BlueTooth. It is a bit more difficult to operate. Also I have a Macintosh; there is a program in that called iPhoto. You can store it there and make albums . . . I can carry my phone and the camera at the same time; it is really convenient. I hardly ever use my digital camera. I always bring it with me in my luggage [on a holiday] but I never really [use it] . . . it is little bit heavy and bulky. It's handy I can always take a snap whenever I am inclined to . . . and the [picture] quality is not too bad. I think quality-wise it is equivalent to a cheap digital camera.
>
> (Australian man, early 30s, interviewed in Istanbul)

And a little later he confesses:

> It is a little bit silly but you know you carry a camera in its case you are immediately like tourists: there is a tourist, you are a tourist, I am a tourist. I like the fact that this just goes to my pocket. I don't have to carry it in its case.
>
> (Australian man, early 30s, interviewed in Istanbul)

This male media researcher has a preference for his camera phone because it is conveniently always at hand, connects well with his computer and blogs and is supposedly less touristic than digital cameras.

Although few Western tourists use their mobile as a photographic device, the opposite is the case with Turkish tourists, both young and old, men and women. Perhaps the most obvious reason for this is cost. Digital cameras are expensive technologies whereas camera phones are cheap when signing a contract (that may make it expensive in the long run). Digital cameras may be out of the reach of less affluent tourists, but 'cheap' mobiles make it possible for them to photograph cost free (see Figure 7.2).

Most digital cameras and mobiles can also record 'moving images', and our observations and interviews indicate that tourists occasionally film to capture soundscapes, funny episodes and the wholeness of places. The woman pictured in the top of Figure 7.3 films for several minutes – slowly moving the camera phone from side to side – to record a Turkish band performing at the Topkapı Palace. A picture cannot capture the soundscapes of this event, and indeed most tourists film rather than photograph during this music performance. The problem of capturing the wholeness of vast places and views with single photographs explains why tourists are keen to scan landscapes by meticulously moving the recording camera slowly from one side of the body to the other, to film the wholeness of a view or place. Such

Figure 7.3 Video photographers.

scanning-in-filming is particularly observed at castle of Alanya with its breathtaking wide views that cannot be consumed with a single static view. As a consequence of the relative popularity of using cameras as filming devices, few tourists today 'sight-see' with both a camera *and* a video-camera, and video-cameras – even digital ones – are rare but not yet extinct (see bottom photo of Figure 7.3).

Novel performances

Elsewhere we have argued that tourist photography revolves around significant places and significant others (Haldrup and Larsen, 2003; see also Hallman and Penbow, 2007) and the present ethnography indicates that digital photography transforms not so much what tourists' picture as *how* they picture, i.e. how they perform photography, as we now document below.

Playful shooting

Both our observations and interviews reveal that tourists take far more photographs with digital cameras than they used to with analogue cameras. This is partly because each click on the shutter-button is cost free and produces a photograph that is erasable (see later). This is in contrast to analogue photography, when each 'click' irreversibly materializes as a material object when the film is handed in for 'pricey' development. So tourists had to worry about cost, running out of film and later development each time a photograph was taken:

> I take many photos today, which is what you do with a digital camera, because you don't have to stick with them. You can always delete them. You used to use more time on each photograph. In the past you didn't just take three or four photographs of the same motif . . . you planned the shooting event and tried you best at each occasion to get the perfect photo. Yes, because the films had only 24 or 36 photographs, so there were limitations.
>
> (Danish man, 60s, interviewed in Alanya)

The interviewees praise this freedom to 'shoot around', and they inscribe it with creative and playful qualities:

> I have an analogue camera . . . but it is a little big to travel around with. And now I have borrowed this digital camera . . . and it has turned me into a photography freak taking 10 photos of everything that I picture, so I now can't use my analogue camera. I prefer the analogue camera, but I like the idea that you can see the pictures immediately, to see if I want a picture of that, if it is good or bad. I'm not particularly brilliant at taking photos so I like the idea that you can see the photos straight away.
>
> (Danish woman, early 20s, interviewed in Istanbul)

Her boyfriend reflects upon why they return home with so many photographs (despite frequently deleting on the move, see later):

> We would never have taken 700 photos, if we had an ordinary [analogue] camera. And this, I think, is the cool thing about it [the digital camera], that you can play with it. There is probably only a few left, but at one time in a hotel I took all sorts of crazy photos of myself in a mirror and it was really good fun. You can experiment much more . . . and this is partly because most of them will not be developed [or printed]. Most of it is rubbish – you are not very critical of the motifs. Of course you try to do a good job and make interesting compositions, otherwise you won't get good photos, but you experiment much more . . . It is a great toy.
>
> (Danish man, early 20s, interviewed in Istanbul)

Other respondents also stress the enjoyment of photographing with digital cameras:

> I love them. I think it is great. That you can actually snap snap delete delete if we need to rather than carrying film and not knowing how they turn out.
>
> (Australian woman, late 20s, interviewed in Istanbul)

These quotations highlight how digital cameras afford experimentation, creativity and play, and how tourist photographers actually use these affordances in practice. As argued in Chapter 1 and above, performance studies argues that performances have both ritualistic and playful elements, and it seems to be the case that digital cameras are pushing photography in a more *playful* direction: they are great toys, a pleasure to play with. The fact that it is free to photograph with digital cameras in part affords such playfulness: 'If I had an ordinary camera I would not have taken 10 films. I mean first you have to pay £5 for the film and then another £5 for the development, so that would be £100' (female student, early 20s, interviewed in Istanbul). But the quotations above also show how the screen, the immediate production and possibilities for deletion reinforce this playfulness, as we will now discuss (Figure 7.4).

Our ethnography indicates how quickly it has become a ritual to examine the screen in the same movement as the shutter-button is released, after a single shot or a a longer series, at the very scene or somewhere with a bit of shade, so that the screen can be seen properly: 'Here you just take five [photos of everything] and then you can sit in the shadow and say: "this is crap, this is crap" and then there is two left . . . there is so much *freedom* involved in this' (Danish female student, mid-20s, interviewed in Istanbul).

As demonstrated in Figure 7.5, one striking finding of our ethnography is the degree to which performances of digital photography are *collective*, or co-performances. Although directing and posing always have been collective performances (as demonstrated ethnographically by Larsen, 2005), *photographing* is now often also performed through collective work, and it has become a ritual to

Figure 7.4 Consuming the screen.

gather around the screen after the shooting to see and eventually delete the image together. The vision of analogue photography is single-eyed and individualized, and onlookers cannot participate in the picturing. Now it is common for co-travellers to participate in the shooting by discreetly looking over the photographer's shoulder, or by giving advice, and thus the eye of the camera multiplies. This also means that women are more involved in their partners' photographing than used to be case.

Figure 7.5 Screens and co-performances.

The interviews indicate two different – albeit often overlapping – rationales for such immediate examination of the screen: production (*inspection*) and consumption (*play*). By production (*inspection*) we mean practices of *evaluating* whether or not a photograph has turned out to be satisfactory, so that one can either move on to the next motif or retake it, perhaps from a different angle, on a different setting or focus, with or without flash, or with different choreography. For instance, in a dimly lit place such as the Blue Mosque, it is challenging to take good photographs; low light requires a long exposure, and the smallest movement of the hand results in blurred or grainy images. Thus, the best-equipped tourists bring out their tripods and self-timer cable releases, but most find support on fences or friends' shoulders or lie on the floor or lean against a doorway. Even then it is difficult to take satisfactory photographs so 'teams' spend much time *evaluating* screens and *retaking* motifs.

With *consumption* (play) we mean the thrill and enjoyment of consuming images immediately. In other words, it emphasizes the *play* aspects of photography. And, if the 'team' retakes a photograph, it is not because they are eager to improve upon it but because it was fun making *and* consuming it. Practices here are not about artful creativity and producing aesthetically pleasing photographs but about the joyful sociability that 'teams' have when playing for and with the camera (see later).

Figure 7.6 is a good illustration of teamwork and *inspection* that shows two men's sustained work (lasting some five minutes) in photographing the information signboards at the entrance at the Alanya Castle. Although only one of them operates the camera, the other one is actively participating, looking through the screen over the photographer's shoulder and giving verbal advice; and after each picture they *inspect* the screen together, to discuss whether the result is satisfactory or if they should retake the photograph, and, if the latter, how they can improve it based upon the previous ones. Figure 7.6 also illustrates how much practical problemsolving in tourism photography happens through collaboration, something also demonstrated in Barry Brown's recent ethnography of map-finding (2007). The practical problem they attempt to solve is to make the signboards readable in image form, something difficult in high sun and if one's camera does not have a special setting for texts. They solve this problem by *inspecting* each photograph and move on when reassured that a satisfactory photograph has materialized on the screen, an assurance that an analogue camera could never deliver – no matter how many photographs one took.

In contrast to the 'serious' inspection and practical problem solving on display in Figure 7.6, the next two figures also highlight co-performances of playful consumption. As ethnomethodologists teach, talk is performed through turntaking, and this is also the case with much (especially digital) photography, when people often shift between being photographers and posers during the same photo session. Figure 7.7 is a good illustration of such photographic turn-taking. Here we witness two young women photographing each other in turn, and once they have taken the photograph they pass on the camera for immediate inspection and consumption.

Figure 7.8 is another illustration of such turn-taking. Here it takes place within a larger 'team' presenting themselves for several minutes against the Blue Mosque. They catch our attention by 'grouping' around the camera after the first photo. At least one appears not to be happy with the result as the 'pose' is retaken, a pose characterized by slightly forward pushed chest and resting hands on the hips. The photograph is then immediately consumed and a change of roles take place. Then a third woman enters the stage as a 'poser', and they meet on half-way for consumption. This time they retake the scene and the image is consumed in close proximity. Then a fourth woman takes on the role of photographing the three others. Again we witness how entangled bodies inspect the image. The following images show the same recurrent turn-taking of roles and patterns of inspecting and consumption. As the event seems to come to an end, it re-energizes when two souvenir sellers approach them and swiftly they photograph each other with 'Turkish' hats. The last photograph shows how this performances ends with group photograph of all the four hat-wearing women taken by their male friend, who until now had been in the background.

Deletion

Although most tourists examine the screen immediately, the interviews reveal different attitudes to and practices of deleting photographs while holidaying. Some people always inspect the screen straight after taking a photograph and they delete those that do not charm instantly. Such 'teams' take far more photographs at attractions than they leave with (and they also delete regularly while they travel and at home):

> We take many more photographs now. This is for sure. And there are many that die as we go by because they are too poor.
> (Danish woman, middle-aged, interviewed in Istanbul with partner)

> I delete them straight off. That's the beauty of digital photography, isn't? I take a bad one I get rid of them and sometimes at the end of the day I go through them as well.
> (Australian woman, late 20s, interviewed in Istanbul with friends)

A second group primarily deletes in the evening in their hotel or when they have 'time to kill', so they have better time and light to judge the photographs. They do so partly to get rid of poor images straight away and partly to ensure that there is enough memory space for the next day's shooting: 'We will have to cut bad pictures as we only have a 512 GB card' (Norwegian man, early 20s, interviewed in Alanya with friend).

Figure 7.6 Co-performing the sign-board.

What are going to do it is to take as many photographs as possible until it is filled up and when we will delete the poor ones. We have only been here three days so it is not full yet but we will do it in the evening.

> (English man, early 20s, interviewed in Istanbul with partner)

Then there is a third group that delete first only when they return home. They stress that they have no urgent need for deleting since they travel with one or more sufficient memory card(s): 'I don't delete, I do it at home . . . the card inside is a big one . . . 500 hundred pictures' (Dutch photographer, early 50s, interviewed in Alanya with partner). Although most tourists delete photographs based upon their appearance on the camera screen, some of the respondents say that they

Figure 7.6 Co-performing the sign-board (continued).

are slightly nervous about trusting the small camera screen (especially if they are outdoors in the sunshine), and thus prefer to delete photographs on computer screens. To quote the female medical student again:

> Hmmm, most of them will not [be deleted on the move]. This is because I have been on a journey where I realized when I came home that some of the photographs that I thought were good were in fact blurred while some of those that I thought were blurred were in reality perfect.
>
> (Danish woman, early 20s, interviewed in Istanbul)

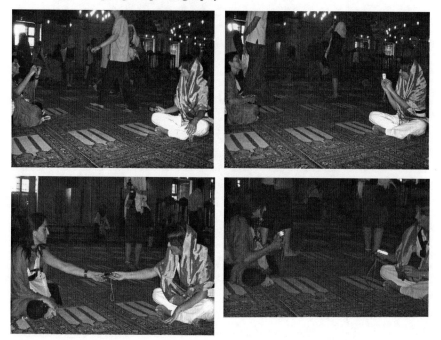

Figure 7.7 Photographic turn-taking.

Future changes

Our observations indicate that the number of tourists using camera phones increased from 2006 to 2007 although camera phone users still constitute only a small proportion of tourists, primarily teenagers and 'stylish' young adults. And yet we have seen how most Turkish tourists photograph with their 'cheap' mobiles. We have also reported that relatively few tourists send live postcards through MMS. Thus, at present the connectivity and mobility of tourist photographs is low, the lack of a critical mass of people with the necessary technology, poor picture quality and the cost of sending MMS messages (compared with SMS) being the major reasons.

But this is likely to change in the near future. The quality of camera phones is improving rapidly, and most new standard phones now come with a 2-megapixel camera – which is enough to take decent photographs. In fact, some of the latest models incorporate cameras up to 5 megapixels (the quality of most standard digital cameras), and Samsung has recently launched a 10-megapixel camera phone. Thus, the perception that camera phones take poor-quality photographs is rapidly disappearing, which also means that the distinction between the tourist digital camera and the everyday camera phone will also disappear. In the near future the camera phone will not only be the camera of the everyday. It seems plausible to predict that camera phones in great numbers will soon replace digital cameras as the holiday camera. Why bring along an additional camera if one has a camera of roughly similar quality in one's mobile phone? The problem with the digital

Figure 7.8 Poses and turn-taking (continued on next page).

camera is that it is one-dimensional, whereas the camera phone is a multimedia device that is always at hand. Soon we might forget to pack or bring out our camera when travelling. And once the cost of sending MMS messages (especially across national borders) and uploading images to the internet is reduced and all mobiles can send and receive them, photo-messages or 'live postcards' are likely to be much more popular.

But there is a possible counter-scenario to the 'death of the camera', this one-dimensional picturing device. Tourists, at least at certain times during their holiday, crave an escape from the perpetual connectivity typifying their everyday life. Sometimes we experience this perpetual connectivity as a tyranny, linked to what Eriksen (2001) calls 'the tyranny of the moment'. In one interview we ask

Figure 7.8 Poses and turn-taking (continued).

a couple who use their newly purchased 5-megapixel camera phone for everyday photography if they will leave their digital camera behind on their next journey. They reply:

Woman: I think is has something to do with tradition.
Man: There is definitely a little element of tradition involved. You have
             to bring a camera with you on a holiday.
Woman: . . . which is crazy because it takes up so much space.
Man: Yes, they do, one could just as well walk around with one's mobile,
             but if I walk around with mobile, I think will also start texting. And

Figure 7.8 Poses and turn-taking (continued).

I don't want to do that when I'm on a holiday . . . that's the rule when we travel, it is unacceptable. It is a relaxation holiday and we don't want any interference.

So, if tourists want to escape the connectivity of the mobile phone, they may take pictures with their camera rather than with their mobile phone. This couple also reveal how it is a deep-rooted tradition to bring along and photograph with a camera, and it may take some time before the majority of tourists regard their mobile as a proper tourist camera, even when they take equally good photographs. This present ethnography has shown that 'screen-ness' is a very crucial feature of digital cameras, and one current problem with most camera phones compared with digital cameras is that they have smaller screens. One important design implication of this is that camera phones need bigger screens in order to become popular. The design challenge involved in this is, of course, how to achieve this without making mobiles too bulky.

Conclusion

Following on from the performance turn outlined in Chapter 1, the first part of this chapter has outlined how we can theorize and research tourist photography as a performance. The second part used such a performance lens to explore ethnographically how and to what extent performances of tourist photography change character and meaning with the digitization of the technologies and objects of photography.

We developed the metaphorical notion of photography *as* performance by drawing upon and extending the perspectives of actor network theory, non-representational theory, performance studies and Goffman's dramaturgical sociology. A performative approach allows the study of photography to move beyond questions and dualisms of representation, on the one hand, and materiality, on the other, without neglecting either of them. A performance approach is concerned with how photography takes place and the embodied, social and hybridized work involved in doing photography here and now, at the 'scene' or at a distance. Following on from Chapter 2 we have argued that ethnography is particularly suitable because it allows lively observations of how digital photography takes place, *as* performances.

We have described how the technologies and affordances of (tourist) photography have changed dramatically in recent years, during which time photography has gone digital and become networked, converging with various other space-annihilating communications and 'communities' such as mobile phones, computers, the internet, blogs and social networking sites. We have argued that the newness of digital photography compared with analogue photography relates to connectivity and mobility, on the one hand (uploading and posting images), and immediate consumption and we might call 'screen-ness', on the other.

Our ethnography has explored how and to what extent the affordances of digital photography are used in practice and thus change the nature of tourism photography. Throughout the chapter we have insisted that digital photography is partly being constructed through being performed. We began by noting how the vast majority of western tourists today (2006–2007) take photographs with digital cameras but surprisingly few use their mobiles for 'serious' tourist photography (they might take an odd mobile picture now and then), because they believe that their mobiles take photographs that are of poor quality. But we have also predicted that this is likely to change in the new future.

We have also shown that few package tourists use the internet to send and upload photographs while on holiday even though there is easy access to the internet at hotels and internet cafés. The exceptions are teenagers and 'round the world travellers'. The major barrier to this is that it requires that tourists bring along the necessary cables and connect their camera to the internet café computer.

Although tourist photography at present involves little 'connectivity' and 'mobility', we have shown the huge significance of the *screen* and how it fundamentally changes the nature of tourist photography, in terms of time, style and sociality. The 'magic' of digital cameras is that they make photographs and they do so instantly, so photographs are widely consumed on the screen and erased on the spot. The quality and especially the size of the screen is pivotal for tourists, and many complained that screens are too small (here digital cameras generally have an advantage over camera phones). Consuming and deleting photographs have become part of producing photographs. These affordances of 'instantaneousness' and flexibility are widely conceived of as pleasurable, and tourists photograph in more causal, experimental and lavish ways as there is no concern about the cost of films and development or about taking bad photographs, as these can always be erased. Doing tourist photography is simply greater fun with digital cameras than with analogue cameras. And a crucial part of this pleasure is the co-performances of much digital photography. Co-travellers constantly participate by looking over photographers' shoulders, and it has become a ritual that the team consumes and inspects the image in unison straight after its release and during the holiday. This is a source of much playful face-to-face talk and sociality. This screen-ness separates digital photography from analogue photography and yet it also highlights their common history. Both the photographic album and the digital camera screen are material objects that function by virtue of presence and co-performances of talk and showing. Yet where we used to come together to relive memories, photographic consumption has now become part of tourists' co-performative experiences.

8   The afterlife of tourism

Introduction

Tourist experiences are often assumed to take place within particular times and spaces detached from people's everyday lives. However, as demonstrated in the previous chapters, most tourist performances intersect with the everyday. In Chapters 6 and 7, we uncovered the habitual and mundane character of tourist performances, showing how embodied 'home-brought' roles, norms, habits, imaginations, dreams and fears are drawn upon and enacted at 'tourist places'.

Building on our method of mobile tourism ethnographies, this chapter takes us into the domestic spaces of tourists. This is done by examining 'how people materialize their travel experiences by following products beyond their points of sale [or production] to explore how consumers appropriate and transform them into a personal effect in the practice of everyday life' (Morgan and Pritchard, 2005: 44). We 'track' – what Marcus and Lash and Lury call 'follow the thing' (see Chapter 3) – the spatialities and temporalities, the physical and digital (im)mobilities of *souvenirs*, i.e. souvenirs' mobile cultural biographies or social life through space and over time. So we explore the afterlife of tourism by examining the souvenirs that our respondents *bought* (e.g. consumer goods), *collected* (e.g. free maps and stones on the beach) and *produced* (e.g. photographs) while holidaying in Turkey and Egypt, and how they enter and are '(re)placed' within and beyond their 'networked homes': used, lived with, reworked, discarded, exhibited and distributed (although we do not examine gift-sharing much). We follow Appadurai's (1986: 5) tenet that 'even though from a *theoretical* point of view, human actors encode things with significance, from a methodological point of view, it is the things-in-motions that illuminate their human and social context'. It follows that the meaning and locality of an object are not fixed, culturally or geographically, but on the move, as it is part of dynamic processes of production, exchange, usage and meaning (see Chapter 3).

We have emphasized that tourism is a 'doing' and tourism performances involve (often pleasurable) 'work'. This chapter examines what tourists do to their souvenirs, to make them prompt memories and be exhibition-able. What we call 'souvenir work' has been neglected in the souvenir and photography literature (but see G. Rose, 2003; Kirk *et al.*, 2006, with regard to the latter). And this notion highlights how individuals actively become producers rather than simply consumers of goods (see also Chapter 1). In a more than representational fashion, we understand souvenirs not only as texts but also as objects. This is in contrast to the (sparse) literature that reads the everyday presence of souvenirs through representational lenses, accentuating their (often trivial) symbolic and textual qualities, as 'signs of tourist travels' (Shenkav-Keller, 1993; Hitchcock, 2000: 2). By doing this, the ambiguous uses of and emotions invested in souvenirs are ignored. To observe souvenir work, we need to depart from 'etic' readings of tourist cultures 'from without' and attempt an 'emic' reading of tourist's home cultures 'from within' (O'Connor, 2006). What interests us is not the triviality of souvenirs or their functioning in abstract systems of signs and symbols, but the personal and emotional investments in and active uses of souvenirs as part of their everyday life over time.

Developing our arguments on materiality in Chapter 4, we show that materiality plays a crucial role in affording memories and imaginative travel, often in unpredictable and unconscious ways. We also highlight how the *mobile* cultural biographies of souvenirs in part are determined by their materiality as some materials travel better and faster than others. And some souvenirs have complex biographies, because they can materialize, dematerialize and rematerialize, take and retake various forms and inhabit different materialities over time. The digital photograph is a telling example thereof. A holiday photograph may well materialize as a photographic object or live a virtual life on the internet; and, yet, someone can always (re)materialize it by making a print from the internet. There has been a recent 'material turn' in photography studies concerned with photographs as material *objects*. This chapter contributes to it (and partly moves beyond it) by discussing digital photographs as *im*material objects *and* the many material moorings that facilitate digital photographs.

The home of tourist souvenirs is tourists' private home. The private home is perhaps *the* single most significant material and affective space for performing the everyday. It is 'invested with meanings, emotions, experiences and relationships that lie at the heart of human life' (Blunt, 2004; Blunt and Varley, 2004: 3). The domestic order is one of inward dwelling and reflection. Moreover, the home is a central space for material and virtual consumption (Miller, 2001c: 3; Chapter 3). Yet now that home has become a 'communication hub', of local and global, personal and public images, a cosmopolitan connectivity stretches it beyond its closed doors and erected walls:

> most of what matters to people is happening behind the closed doors of their private sphere. The home itself has become the site of their relationships and their loneliness: the site of their broadest encounters with the world through television and the Internet, but also the place where they reflect upon and face up to themselves away from others.
>
> (Miller, 2001c: 1)

Not only is the networked home the place where we daily travel the world on various networked screens (and plan and book our journeys), it is also where souvenirs are worked upon, displayed and circulated (and made available for future holiday planning). Souvenirs are among those personal building stones that both turn a house into a *personal* home and stage a desired self-identity in the eyes of and needing approval from guests (objects as Goffmanian props).

During the home interviews (see Chapter 3) people showed us around their homes, pointing out where and how they store, display, circulate and use their souvenirs. We looked at and talked about souvenirs contained in and exhibited upon walls, frames, fridges, photo albums, shoe boxes, jewellery boxes, digital photo albums, mobile phones, cameras, computers, VHS tapes, CD-ROMs, DVDs, email accounts, blogs, social networking sites, and so on. So we moved around in their houses while interviewing, and conversations often revolved around objects or screens at hand. Together with our respondents, we 'excavate' some of the multiple ways in which the performances of tourism integrate with their everyday domestic spaces.

Ambiguous souvenirs

Trivial souvenirs

In academic literature, the tourist souvenir, whether we speak of craftwork, mass-produced consumer goods or personal photographs, is generally treated as the very epitome of (tourism's) kitsch. Although tourists presumably look for local culture, their souvenirs are collections of 'genuine fakes' (Shiner, 1994; D. Brown, 1996). They are tacky objects superficially referring to places visited by the owner. Paradoxically, trivial signs and markers are used as Goffmanian props to display adventure, edification and global outlook. But rather than local culture and global outlook, the souvenir represents frictionless mobility and results in a sort of 'geographical displacement' of objects' authenticity as when tourist shops at the Niagara Falls sell Egyptian-themed souvenirs (Hashimoto and Telfer, 2007: 204). As O'Connor observes, early tourism theories weigh heavily on contemporary studies, and souvenirs seem to fall victim to the former's dramaturgical vocabulary of alienation and inauthenticity: 'when material objects do figure [in contemporary studies], they are treated *a priori* as kitsch' (O'Connor, 2006: 253). Hence, critical studies (un)critically reproduce the infamous authenticity and inauthenticity dichotomy. As Goss says:

> [I]f tourism is conceived as a quest for authenticity in the world of the Other
> . . . the gullible tourist is presented with only 'reconstructed authenticity' and
> staged 'authenticity' behind which lies a hierarchy of 'backstage' regions, or
> reserves of more or less authentic culture to which they have no access.
>
> (Goss, 2004: 330)

In writings on personal/tourist photography, we also discover contempt and ridicule. Don Slater (1995: 134) remarks that 'snapshot and amateur photography are generally regarded as a great wasteland of trite and banal self-representation. Much of this limitation rounds on processes of idealisation of the self and the family'. Like tourism, personal photography produces a small world of positive extraordinariness. This ties into Pierre Bourdieu's statement, 'the photograph itself is usually nothing but the group's image of its own integration' (Bourdieu, 1990: 26). And the snapshot world is said to be inhabited by identical family albums (Chalfen, 1987: 142; Halle, 1991: 104). The literature on tourist photography portrays this emblematic tourist performance as a wasteland of pre-programmed shooting of image-driven attractions. Here, tourists are not so much framing as already framed by the tourism industry's economy of signs that form a 'hermeneutic circle'; so tourists return home with identical photographs that, in turn, mimic the commercial images they had already consumed at home (Sontag, 1978; Urry, 1990; Osborne, 2000).

The above illustrates that aloof and critical readings of souvenirs inescapably arrive at the conclusion that they epitomize staged authenticity, the kitschy and the trivial. There is a certain veracity to such a reading. For instance, the markets in Egypt and Turkey where we did research and our respondents (and we!) purchased souvenirs fit this critical description in many ways. The goods here are an uncanny mixture of supposedly 'authentic' *local* stuff and *global* brands unified within an atmosphere of the 'genuine fake' and 'global compression' regardless of their geographical 'belonging'. The 'local' goods are largely mass-produced tourist kitsch, whearas expensive fashion – clothing, perfume and sunglasses – brands such as Armani, Dolce and Gabbana, Versace, Louis Vuitton, Lacoste, Chanel, Ralph Lauren, Bjørn Borg, Clavin Klein, Adidas, Puma and Nike are (largely) cheap unlicensed *copy* products. In addition to such global fashion brands, the jerseys of Europe's most famous football clubs, e.g. Manchester United, Liverpool, Arsenal, Chelsea, Inter Milan, Bayern Munich, Barcelona and Real Madrid, are sold everywhere. The brand of Hollywood is equally on sale with an abundance of Spiderman, Batman, Shrek and Bob the Builder t-shirts, caps, rucksacks, bedclothes and much more.

These global brands are sold everywhere around the world, and they have no particular relation to Egypt and Turkey (except that many of their *original* products are made in Turkey). Such global brands can be said to be placeless, mobile souvenirs not bound to any particular place, and they seem 'in place' everywhere; their only mooring seems to be a culture of global consumerism. Such global brands are clearly first and foremost not souvenirs of Turkey and Egypt (or any other place) but souvenirs of 'global modernity'. As our casual conversations and formal interviews reveal, tourists buy them because they are much *cheaper* – being 'copies' – than at home: the brand value and the cheap price is the attraction. Low cost also explains why many buy gold jewellery in Turkey; indeed, for repeat tourists affordable gold jewellery is one attraction of Turkey. Few buy ornamental Oriental-style jewellery, preferring minimal Danish-style jewellery fitting their everyday wardrobe. The popularity of 'expensive' fashion brands and silver/gold jewellery indicates 'conspicuous' consumption, and the symbolic and displaying use of ('disguised') souvenirs at home.

In some of the homes we visited, souvenirs play invisible and few roles precisely because our interviewees regard souvenirs as trivial and worthless. Such people would at best, when asked, 'retrieve' souvenirs hidden in cupboards, store rooms or cabins or underneath beds or stored on computers. Souvenirs do not 'belong' in the living room; they do not merit visual or public attention in the front-stage regions. They can *only* be displayed without voicing their 'origin' of tourist travel too loud; souvenirs need to disguise themselves. The wooden figure in Ann-Lee and Leigh's (home ethnography 6) house has been pursued and displayed because 'it is nice, such a fine piece of craftsmanship' and the alabaster jar in the windowsill 'glows with such a beautiful, warm orange-red glow when you put a candle in it'. As Leigh explains more generally:

Leigh: But . . . Ann-Lee, we didn't buy anything at all in Egypt to bring with us home, did we?
Interviewer: Nor can I see any! Why not?
Leigh: Because we do not want all that *shit to dust*!

(home ethnography 6)

For such people, souvenirs are generally regarded as 'shit' that eventually will 'haunt' – like 'dust'. Or souvenirs are regarded too 'cheap' or too 'colourful' to be at home within a modernistic white-wall home where simplicity rules. Here, souvenirs are allowed to 'inhabit' only the stylish Macintosh computer (Figure 8.1). In some homes, the only type of souvenirs that are 'in place' are personal holiday photographs – in part because they are manageable in albums and databases. And, similarly, not all objects of travel are wearable, capable of belonging on tourists' bodies at home. Lotte explains:

Lotte: Because of the wind, we bought some scarves . . . we got some from the natives who wrapped them around us, you know, around the mouth and the nose, you were able to see . . . but you do not get sand [in the eyes] . . . it was only because it was practical . . . But it is actually quite funny because it is actually fashionable now . . . I've got a scarf here that is worth I do not know how much in the [high street] shops down here.
Interviewer: Do you then use it from time to time?
Lotte: No. It's simply not my style . . . but I intend to bring it with me down there again.

(home ethnography 11)

Meaningful souvenirs

Critical readings of or attitudes towards souvenirs, such as the ones expressed above, invoke a paradox. Despite all their triviality and fakeness, one can argue,

Figure 8.1 Stylish screen.

and our home ethnographies indicate this, that *some* particular souvenirs (not only 'disguised souvenirs') may well be meaningful, valuable and usable to the owners, not necessarily for good, but at least for some time. Although global brand souvenirs (e.g. a Barcelona or Spiderman t-shirt) do not signify Turkey-ness or Egypt-ness for the owners, they may nonetheless invoke strong memories of these specific places. In this sense, a global brand souvenir can become a souvenir of a particular place, event or situation (and this is precisely why it is a souvenir!). For instance, when Jonas's son, Elliot, wears his 'Bob the Builder' t-shirt bought in Alanya, Jonas instantly recalls Elliot's smile when he proudly paraded it at the hotel for the first time.

Much the same can be said of holiday photographs stored in shoe boxes or albums and now also on databases. Although they are deeply conventional, we feel an extreme affection for them. It is, after all, our (former!) 'loved ones', our children, our holidays and our life stories. As discussed in *Performing Tourist Places* (Bærenholdt *et al.*, 2004: Ch. 6), Barthes (2000) mournfully demonstrates that personal photographs affect us by sentimentality – they are affective objects of love and death, they are an 'order of loving'. People dream, remember, hope, despair, mourn, gossip, hate and love with their photographs. And this can also be the case with purchased or collected souvenirs, if they contain and trigger particular personal stories and sentiments for the owner. In her book *On Longing*, Susan Stewart (1993: 136) notes that the narrative enabled by the souvenir 'is not a narrative of the object; it is a narrative of the possessor'. The crucial point is that such embodied narratives will be (momentarily) available only to the owner of the object or image – never to the aloof researcher. People engage with, and 'make sense' of, their souvenirs not through detached semiotic readings, but expressively, through the 'feeling body'. This insight partly prompted Barthes' phenomenological writing on photography in *Camera Lucida*:

> I realized with irritation that none discussed precisely the photographs which interest me, which give me pleasure or emotion. What did I care about the roles of composition of the photographic landscape, or, at the other end, about the Photograph as family rite? . . . Myself, I saw only the referent, the desired object, the beloved body . . . I wanted to be primitive, without culture . . . So I make myself the measure of photographic 'knowledge'. What does my body know of Photography?
>
> (Barthes, 2000: 7, 9)

Here Barthes theorizes photography with, and through, his own lived body and a private photograph of his newly diseased mother. This has affinities with Douglas's argument that 'we may often be on the wrong track trying to decide what [objects] signify, since that question does not necessarily lead directly to the part the objects play in human transactions' (cited in G. Rose, 2003: 7).

Moreover, rather than being an alienating add-on, souvenirs are an integral component in producing identity, social relations and 'familyness'. We may even propose that 'people require [tourist] adventures in order for satisfactory life stories to be constructed and maintained' (Scheibe, 1986: 131). Through buying, producing and displaying souvenirs, people strive to make fleeting tourist experiences a lasting part of their everyday life and life narrative.

Souvenirs are not only relicts of our spatial trajectories in a distant past, but also important elements for stabilizing our 'sticky' everyday life through our personal material cultures. When doing souvenir work, families are actually in the process of making their experiences and performing 'familyness'; it is part of that theatre that they enact to produce and exhibit family narratives and memories. In this way, souvenirs may be read as 'transporters', intermediaries connecting spaces of tourism and the everyday. Souvenirs populating and displaying our bodies, living rooms, studies and kitchens act as 'the set and props on the theatrical stage of our lives' (Wallendorf and Arnould, 1988: 531). They do not only help to express and establish 'where we were'. They must be read as 'touchstones of memory', tools for our own personal, dynamic processes of becoming (Morgan and Pritchard, 2005: 31, 41; Digby, 2006). They continually afford establishing who we are (or want to be) and with whom we are (or were), whereas souvenirs put away express our inability (or unwillingness) to live with them – the old stories that they still trigger or new ones that they now prompt. And yet tourists who fail to impose 'domestic order' over their souvenirs face the danger of 'clutter' within their living spaces. Souvenirs cluttered in the living room, stored in the cellar, given away or discarded may still live on and induce their 'absent presences' on the lives of people. Souvenirs can 'haunt' the lives of the living: 'Things we threw out before we should, things we held onto long after they should have been disposed of, the credit card statement at the end of the month' (Cwerner and Metcalfe, 2003; Hetherington, 2004: 170; Gregson *et al.*, 2007).

As material objects, foodstuff, clothes, furniture, artwork, jewellery, and so on, transported home as souvenirs may encapsulate – in addition to their use-value – the atmosphere of a particular tourist place as well as embodied memories of spicy food, the warm sun, the smells of the harbour, the sounds of the bazaar, and so on. Such apparently trivial objects can turn into supernatural mementos when brought home as souvenirs. Thus, it is not as a mere sign, vaguely reflecting a weak afterglow of the 'aura' of places and events, that souvenirs achieve phantas-magoric powers. The value of the souvenir rests on its ability to transform into a 'trace', 'a manifestation of closeness, however distanced it may be', enabling us to 'enter into the possession of the thing' (Benjamin cited in Markus, 2001: 33; see Bærenholdt and Haldrup, 2004: 86 on the difference between 'aura' and 'trace'). However distanciated we may be from the places, events and people we visited and enjoyed, the souvenir affords closeness with them. It is as a material remainder embodying the possibility of imaginative revisit and re-enactment – mediating between the realms of past and present, materiality and meaning, 'reality' and fantasy – that the souvenir gains its power. The souvenir presents what is absent; it 'domesticates on the level of its operation: external experience is internalized; the beast is taken home' (Stewart, 1993: 134). As material objects that hold on to those moments that would otherwise soon wash away, souvenirs – especially photographs – connect with and revive memories through 'memory travel'.

Memory work

Yet, on their own, souvenirs do not produce memories (see Bærenholdt *et al.*, 2004: ch. 6). Although affording memory stories that would not have been prompted without them, they are not receptacles for memories. Memory is a complex of relational interactions between humans and images or objects. We call such interactions memory work. The souvenir sets off the memory journey, but it is hardly the destination. Interactions with souvenirs stimulate contingent memories that can move far beyond what the image depicts or the object is:

Keith: We bought Turkish Delight.
Lisa: Yes we did, for your father and my grandmother, for her birthday. It turned out to be a major hit because my grandfather once gave it to her too as a sweetheart present shortly after they met each other.
(home ethnography 1)

As a souvenir 'tasting of Turkey', Turkish Delight, in the interaction with the grandmother, takes a completely new turn: Turkish Delight now reactivates delightful memories of early love. A given souvenir can prompt many stories and emotions as time pass by and people move through life. The meanings of our souvenirs are seldom static, because our social biographies are characterized by flux and rupture as much as by stasis. Although they are objects and images of the past, souvenirs are always about today. They re-enter the flow of time through people's actual *use* of them. As Anette Kuhn says in relation to family photographs, they:

> effect to show us our past, but what we do with them – how we use them – is really about today, not yesterday. These traces of our former lives are pressed into service in a never-ending process of making, remaking, making sense of, our selves – now.
>
> (Kuhn, 1995: 16)

Death 'haunts' our loving family photographs. Our dead relatives and our once youthful bodies stare at us. Or our former partner – whom we perhaps still love! This is the dark side of photography, of loss, absence and death (G. Rose, 2003). Souvenirs and memories are always about making sense of one's past life and social relations today. And yet it is also the case that photographs especially are precious objects affording 'timeless' journeys along paths traversed many years ago: as a child, a teenager, a newlywed, a father . . . Their value increases with the passing of time (Bærenholdt *et al.*, 2004: ch 6). Souvenirs (especially photographs) are inherently ambivalent emotional objects. The significance of a lovely family album or piece of anniversary jewellery changes completely if love turns into hate.

The life cycle of souvenirs

Souvenirs are ambiguous and dynamic objects, destined to live an unpredictable afterlife. Not all souvenirs live long. Some are pursued for their price and convenience, to be used and disposed of when worn out. As John says:

> I know that I bought copy products. But I don't care. You only pay 400 kr. [£40] for three pairs of jeans. And, well, if they last a year then it is great. You don't expect them to last longer.
>
> (home ethnography 15)

Yet 'durable' souvenirs may also live short lives. Sometimes we discover that a long-pursued souvenir is 'out of place', being perhaps too 'kitschy' or not prompting any memories, or the wrong ones, once it enters our domestic space. In contrast, souvenirs casually 'picked up' without much thought may turn out to travel and inhabit our living room well (Digby, 2006: 171). And souvenirs that did 'fit in' initially, lovingly appreciated and cared for at one time, may perhaps be a burden or shame us tomorrow. They can suddenly 'strike back' with bad memories. Souvenirs once 'in place' can grow uncomfortably 'out of place' or 'out of fashion'; they have to be removed from our sight or relocated to less prominent spaces.

Or they may simply be forgotten by the owner: '[S]ouvenirs help us to create a continuous and personal narrative of our past and herein also lies any souvenir's "tragedy" since it is ultimately destined to be forgotten in the death of memory' (Morgan and Pritchard, 2005: 46). Or precious, long-cared-for souvenirs may break or disappear against our will. The 'irony', as Tolia-Kelly (2004a: 315) says more generally about our everyday material cultures, is that they are the 'essence of security and stability', yet they 'are sometimes transient, ephemeral things which in turn fade, tear, fragment, dissolve and break'. As our preferred narratives may develop, transform and shift, so must our material tourism cultures.

When visiting the home of Ann and her daughter Pennie (home ethnography 2), the 'absent presences' of souvenirs are manifest. Despite a gigantic world map and several travel books, no other souvenirs are visibly on display in the living room. Ann bought many souvenirs in the past, but now they embarrass her. As she says with regard to the souvenirs purchased in Egypt a couple of years ago:

> I think I brought them to document the history of Egypt, statues of Akn-aton, papyrus, that sort of things. I did think they were fantastic from an aesthetic point of view, but I also wanted to show them at home, a kind of documentary: 'see, that's Egypt!' . . . I wished to document it to my family, especially my grandparents who are very local.
>
> (home ethnography 2)

She bought souvenirs to document and mobilize experiences and 'Egypt' into her local life-worlds, so that members of her social network – especially her grandparents – could travel along with her and experience Egypt too. Once they fulfilled that 'guiding' – its use-value – they somehow ceased to be meaningful, and they have become a little embarrassing, almost tacky and certainly very 'touristic': trivial tourist signs. They now live a quiet afterlife on the shelves just below the corridor ceiling; although not discarded or completely put away, they are hardly on display or part of the front-stage (Figure 8.2, top). These souvenirs' cultural biographies, from objects of beauty and documentation to being pointless and causing light shame, encapsulate what we call the mobility and ambiguity of the souvenir.

The souvenir that has made the transition into her everyday life best is a bag made from camel hide that crucially did not look like a Gucci copy and yet resembled a Burberry one. She had seen it at home but could not afford it (Figure 8.2, bottom):

> I missed a bag. Back home I had looked at a Burberry but it cost £500. And then, I found this one that looked a bit like it without being some kind of Gucci-like copy. It should not look like a copy . . . and it was funny that it was made from camel hide . . . and part of the story is that it was very cheap, but then that is always a part of my shopping stories: 'I got this very cheap!'
>
> (home ethnography 2)

What this example illustrates is that valuable souvenirs facilitate personal stories without drowning them with too impersonal, abstract and cheap tales or touristic

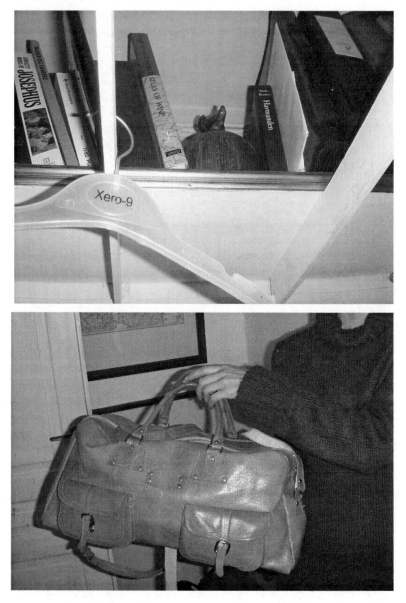

Figure 8.2 Ann's souvenirs.

signs. It also highlights that valuable souvenirs often have use-value, are useful or usable in the everyday.

The souvenirs purchased by one young couple, Lisa and Keith (home ethnography 1), also highlight the ambiguity of the souvenir and illustrate that they need to be 'useful' and disguise some of their tourist origin to be appropriated and set root within the domestic order. Their flat exhibits several holiday souvenirs, including a shisha pipe bought in Turkey, two big Moroccan cushions on the floor, cushions from Latin America and Turkey on the sofa and a blanket from Vietnam (Figure 8.3). Lisa likes the cushions and think they 'fit in' because they are not 'typical,

Figure 8.3 Keith and Lisa's living room.

not Oriental in their style, you cannot really see where they are from . . . but if someone asks I will of course tell them'. Later on she says: 'I think it is so cool that my pillows are from Guatemala and Istanbul and the blanket from Vietnam. Yes I like that' (home ethnography 1). For them, the 'right' sort of souvenirs is a crucial way of presenting a cosmopolitan identity (as discussed in Chapter 9).

They buy souvenirs that are 'typical' of the place they visit without being blatantly touristic and 'cheap' ('it is of course funnier to buy something in Turkey that has a touch of Turkey' (Keith, home ethnography 1)) and, equally important, have *use*-value at home:

Keith:	Everything we have bought is something that we have used. We do not buy any thing just to buy a souvenir.
Lisa:	We have not bought anything just for decoration.
Keith:	. . . we buy things that we can use so that they do not just fill up space.

(home ethnography 1)

Although insisting on buying 'souvenirs with use-value', sooner or the later some of their souvenirs turn out to have little or the wrong use-value at home. Things may belong within the places of tourist performances and yet grow 'out of place' at home. Not all souvenirs travel well and can be appropriated at home. Objects have to belong within certain spaces not to be 'out of place'. Having enjoyed drinking Turkish tea while holidaying *in* Istanbul, they bought Turkish tea for themselves and their family members. Yet, at home, they realized that it was the setting that made the tea taste good. 'Out of place' the tea is 'bloody synthetic', 'sweet, disgusting trash':

Keith:	I also bought some Turkish tea for him [father], which he probably has not used. No, yes, he has used it only to say that he has used it!
Lisa:	It did taste much better in Turkey . . .
Keith:	. . . it actually did taste better in Turkey . . .
Lisa:	. . . it is bloody synthetic.
Keith:	. . . sweet, disgusting trash (ha, ha, ha) that only tastes good when you sit with a view over the Bosphorus Strait.

(home ethnography 1)

Somewhat reluctantly, they also admit that this is also the case with shisha pipe smoking 'out of place':

Keith:	We buy the tobacco in our local grocery shop [Turkish owner]. So it was actually the same that we brought home from Turkey, but it is just a little more funny to buy it there [in Turkey]. We used it a lot in the beginning. Sometimes we even smoked on our own [without guests].
Lisa:	But you get a headache.

Keith:	Yes, I quit smoking and you get a headache too.
Lisa:	Well, yeah, I get a headache too.

(home ethnography 1)

So they very rarely use it these days. The shisha pipe was purchased not only for its aesthetic and 'cosmopolitanism', but also because of the pleasures of smoking that they experienced in Turkey and desired to translate into and re-experience at home. Now, the shisha pipe recalls bad headaches as much as 'smoky Oriental nights' and it is downgraded to a decorative object of little use lurking on top of a bookshelf. Although bought to resurrect 'Arabian nights', it now collects dust rather than filling the apartment with sweet tobacco. Its future within their home seems very uncertain! They also found that flies sometimes fly along with spices, so spices do not always travel well: 'we had to throw it [the spices] out. When we came home we discovered some sickening flies in the spices' (Keith, home ethnography 1).

(Im)materialities of photographs

Before digital photography really became commonplace, Elizabeth Edwards argued that personal photographs are not only two-dimensional images, but also three-dimensional things (something neglected by conventional photography theory), often with a permanent afterlife. She is inspired by Barthes' famous description of the photograph of his deceased mother in *Camera Lucida* (2000):

> The photograph was very old, the corners were blunted from having been pasted in an album, the sepia print had faded, and the picture just managed to show two children standing together at end of a little wooden bridge in a glassed-in conservatory, what was called Winter Garden in those days.
>
> (Barthes cited in Edwards and Hart, 2004: 1)

What Barthes describes here is not so much what the photograph shows but how the passing of time has physically aged it as an object – just as the passing of time aged and eventually took the life of his mother. Just like the human body, photographs exist materially in the world, and they have a social biography over time and through space that marks them. It is not only the content but also the 'body' of the photograph that reveals its age. For instance, 'albums have weight, tactility, they often smell, often of damp, rotten card – the scent of "the past"' (Edwards and Hart, 2004: 11). Whereas the sensuous geography of the photograph as image is purely visual, it is much richer as an interactive and tactile *object*: we can touch and perhaps smell it. We want to argue that this 'aura of thingness' assures photographs a long afterlife. Furthermore, when people engage with photographs, as Edwards observes,

> the describing of the content is accompanied by what appears to be an almost inseparable desire to touch, even stroke and spit on the image. Again the

viewer is brought into bodily contact with the trace of the remembered. Thus we can say that the photograph has always existed, not merely as an image, but in relation to the human body, tactile in experienced time.

(Edwards, 1999: 228)

Photographs have auratic properties due to their material existing in the world: an ' "aura" of thingness' (Edwards and Hart, 2004: 9). This 'aura' is further underscored by their 'inhabitation' of material objects such as albums, frames, shoe boxes, wallets, pinboards, fridges or jewellery and the fact that they are easily passed from hand to hand around a coffee table.

Yet now that most people make photographs with digital cameras, photographs and photographic memories are not always objects; many photographic images are now destined to live, for shorter or longer periods, virtual, digital lives without any material substance – in cameras and computers as well as on the internet. And this lack of an 'aura of thingness' partly explains why so many digital photographs are short-lived; in fact, many are 'aborted' immediately after a swift viewing on the camera screen (as discussed and shown ethnographically in Chapter 7). Our observations and interviews indicate that most digital photographs are consumed on various screens rather than on printed paper. In their early days, they are consumed on the camera or mobile phone screen, and if they survive deletion at this stage they might end up being uploaded to a computer, entered on databases and viewed on yet another screen: the *computer* screen.

Although most of our interviewees had uploaded their digital photographs to their computers, we also visited a few homes where holiday photographs seemed destined to be forever imprisoned by the camera that gave birth to them. Despite having a computer in the bedroom and returning home from Turkey with photographs half a year ago, this couple in their early 20s state:

Morten: They are still in the camera. Even though it is a digital camera I have never transferred them to my computer . . . Nor have we printed them. The only one I think that has seen them is my mother – once.

Camilla: We do not even know where it is; they are in the drawer somewhere.

(home ethnography 12)

This couple was far from the only people who expressed a longing for paper photographs and physical albums. Although most delight in photographing with digital cameras (as demonstrated in Chapter 7), some say that they are less thrilled about consuming holiday photographs on computer screens. Several complained to us, that in comparison with photo albums, computer slide shows are a little too impersonal and too speedy. Moreover, they do not really afford sustained, communal or cosy viewing as people often rub shoulders in studies and bedrooms, where most stationary computers reside.

Once photographs 'inhabit' networked computers, they can travel the world as digital bites of information and put (perhaps fleeting) roots in various email boxes, home pages, blogs and social networking sites. Personal photography came late to the internet. Yet home pages, blogs and, not least, the phenomenal rise of social networking sites within the last couple of years mean that millions of personal photographs are uploaded to the internet on a daily basis. Here, they become part of our (presentation of) distributed virtual selves and life narratives: 'the function of memory reappears in the networked, distributed nature of digital photographs as most images are sent over the wires and end up somewhere in virtual space' (Van Dijck, 2008: 58). Photographic memories are dislocated from the fixed physical home and object-ness, and distributed to the (semi)public spaces of the virtual and the global internet. Photographs can travel much faster and gain a much larger audience in digital form on the internet than in material form by traditional (slow) mail. This digitization, or dematerialization, of photographs and their convergence with the internet has made them truly shareable; internet-connected computers share photographs with great ease (see also Chapter 5).

In Chapter 5, we noted the abundance of Oriental images exhibited by 'ordinary' people on various travel and photo sites. Users of such sites – in particular Flickr – tend to be keen photographers with aesthetic ambitions; for them photography is a 'hobby' (Miller and Edwards, 2007). Such amateur photographers upload between 2.5 million and 3 million new photos on Flickr every day,[1] and in November 2008 the site claimed to host 3 billion photographs (2 billion the year before).[2] Some 80 per cent of Flickr's users exhibit their photographs to a global audience (Davies, 2006; Cox *et al.*, 2008), which means that everyone can 'travel' to them, comment upon them, and link them to other blogs and sites. This is in contrast to much analogue photography, which was introvert and consumed in solitude or face-to face with friends and family members. Our only respondent who uses Flickr fits that type. Henrik (home ethnography 5) is in his mid-40s and a very passionate photographer. He spends a great deal of time editing the photographs that he uploads to Flickr in Photoshop, a professional editing program. His profile is 'open' and is exhibited there because it provides an audience for his photographs of places around the world. There is a certain anti-touristic sentiment to his photographs, as none of them includes friends and family members; they are artful photographs. And in the open kitchen where most of the interview takes place, his 'trophy shoots' are tastefully framed on the wall (Figure 8.4).

In Chapter 3, it was noted that 30 per cent of the Danish population now has a Facebook account compared with only a few hundred thousand (mainly youngish people), or fewer, when we conducted the bulk of our home ethnographies during 2007. Although only a few of our interviewees had a Facebook profile at the time of our visit, many more of them have today. We have become 'Facebook friends' with some of them, and they – like most other Facebook-ers – exhibit single images and albums here (see next section).

Users of Facebook have uploaded more than 10 billion photographs to date, with that number increasing by an astonishing 700 million each month.[3] In contrast to Flickr, most profiles on Facebook are 'private'. So it is only 'friends' – people whom one has *accepted* as one's 'friend' – who have access to one's private photographs. Moreover, whereas most photographs on Flickr are 'arty', on

Figure 8.4 Photographic walls.

Facebook they are more mundane and candid. Unlike the traditional photo album, exhibitions of photographs on Flickr and Facebook tend to reflect 'instantaneous time' and mutability; they are tied into the flow of the everyday. They are photographs of last night's party, the son's first bike ride yesterday, last week's journey, last month's holiday, and so on. Facebook-ers do not so much share memories as they share *experiences* of events that recently took place. However, old analogue photographs occasionally also appear, as when old school photos are scanned, uploaded and distributed to old school friends. Although it is still early days, the lives of photographs on Facebook and Flickr tend to be short-lived. Van House's empirical article concludes that Flickr collections (perhaps less so with Facebook because of their personal content) are a stream of 'transitory, ephemeral, "throwa-way"' images:

> However, while some people are using Flickr to archive their image collections, most participants see Flickr as a social site, a place for *sharing* images. They described their Flickr collections as *transitory, ephemeral, 'throwaway,'* a stream, not an archive. Their primary interest was in recent images, their own and their contacts. Images are archived elsewhere.
>
> (Van House, 2007: 3, our italics)

This quotation also highlights another key feature of Flickr and Facebook: they (are designed to) afford easy *sharing* of photographs to one or all your 'friends'. In this sense, digital photographs are a crucial component of mobile networked societies where one's ties often are at a distance and much socializing is mediated by and conducted through screens at a distance (see Chapter 3). *Sharing* (and 'accessing') photographs and virtual albums on the internet helps to bridge such distances. Gillian Rose (2003: 11) speculates that: 'the more distant people are, the more important photographing becomes . . . A major reason for sending family photographs to relations and friends is that they do not see the children of the family frequently enough'. Rose continues:

> family snaps are seen as a trace of a person's presence; but they are also taken, displayed and circulated in awareness of the pervasiveness of absence and distance. Hence the spatial stretching of domestic space beyond the home. Photos bring near those far away.
>
> (ibid.: 13)

In contrast to traditional emails, Flickr and Facebook allow more than just the discreet sharing of photographs between distanciated 'friends'. Each time one of your 'friends' uploads photographs or is 'tagged' in one of his or her friends' photographs, you and *all* his or her friends are immediately informed about it in their 'news feed', and they are consume-able with one click: it is one extensive network. This also means that you are 'exposed' once you share photographs on Facebook and you never know who consumes them. There is a certain voyeuristic element to such disembodied and faceless browsing through, say, an old school-mate's family photos – people you perhaps have not seen since leaving school some 20-odd years ago! And yet the design of Facebook and Flickr also encourages public viewing and 'photography talk'. Each photograph is accompanied by a comment box where all your friends can write a public comment about any of your photographs. Such 'commenting' is very common on Facebook.

Given that the average Facebook-er has some 100 friends, Facebook-ing has become an integral part of the everyday, and Facebook photographs are much seen and commented upon; personal photographs now reach a much wider audience (including 'weak' and 'old' ties) and have become part of the everyday life of the networked household and its face-to-screen sociality. They have never been so visible, distributed and semipublic before. Jonas's 136 'friends' have, for example, uploaded 291 albums that are automatically filed in one archive (Figure 8.5). Once a photographer uploads a photograph to Facebook (or any other such 'site'), he or she loses control over its destiny, as friends (and even strangers if the profile is a public one) may use it in unforeseen contexts or distribute it even further. As copy-able and timeless travelling bites of information, they face very unpredictable futures with multiple possible paths.

We are faced with a paradox as photographers also gain more performative control over their images. Whereas Kodak once said, 'You press the button, we do the rest', they – and the digital photography industry more broadly – now seem to say, 'We deliver the network, you – through post-photography – *and your networked computer* edit (through "cropping", "rotation", "red eye reduction", "light enhancement", "scene/light balance" and various "fun effects"/"creative project"), organize, print and distribute photographs':

> What is new in digital photography is the increased number of possibilities for reviewing and retouching one's own pictures, first on a small camera screen and later on the screen of a computer . . . In other words, *does image doctoring become an integral element of autobiographical remembering?*

Figure 8.5 Albums on Jonas's Facebook profile.

Digital doctoring of private snapshots is just another stage in the eternal choreography of the (mental and cultural) image repertoires once identified by Roland Barthes.

(Van Dijck, 2008: 67, our italics)

Although a few of our interviewees did basic editing, the majority did little more editing work than 'filing' their holiday photographs, so there is little evidence in our research to support the idea that 'image doctoring' has 'become an integral element of autobiographical remembering'.

Although we have seen that many digital images exist virtually, digital photography is not without a material substance, and some digital images do materialize as photographic objects. Mobile phones, and especially digital cameras, are designed to be networked with material objects such as home computers, printers, toner, CDs, photo paper and photo albums; only then do they potentially realize their (relational) affordances in full. These material objects of networked communication enable images in camera phones and digital cameras to be enlarged and improved aesthetically, and to travel and appear on paper and other material surfaces. Without such more or less mobile and immobile 'moorings' and facilitators, digital photographs would not travel far (MMS messages being the exception) or gain much material presence (Larsen, 2008). Finally, although our home ethnographies indicate that the vast majority of digital images do *not* get printed, many of our respondents still – or at least plan or expect or desire to – print their best or most precious photographs, believing that in this journey from the digital to the material, value – 'aura of thingness' – is added to the photograph. Indeed, Edwards suggests that:

> human values and human desire for linking objects of memory will, I believe, still demand the material possibilities of photography, where the affective tones of physical tactile quality . . . integrally construct the photograph and its status as an object of memory.
>
> (Edwards, 1999: 236)

Inside the home

In order to further illuminate these diverse uses and disuses, presences and absences, joys and embarrassments, im(materialities) and (im)mobilities of purchased and produced souvenirs, we now 'look into' four private homes where we did interviews and observations to tell different stories of the afterlife of souvenirs within and beyond their domestic home.

Although she first embarked on a journey abroad in 2000 (seven years before our visit), when one enters the living room of Susan's (home ethnography 9) small flat, an abundance of souvenirs from around the world meets the eye, speaking of travels to and relations to faraway places because of strong ties at a distance (Figure 8.6). Here, souvenirs have much presence and significance in turning a home into a personal place. Her family is scattered around the world:

Figure 8.6 Susan's living room.

> Actually, I have seven nationalities in my family . . . it started with my
> father's brother moving to Canada . . . because of unhappy love. Then he fell
> in love again and has been over there ever since. My aunt was married to a
> Jew and has settled in Haifa and has been living there for 40 years now, and
> my cousins from down there have married and settled all over the world, and
> then my father has just recently married a woman from Romania.
>
> (home ethnography 9)

We interviewed Susan because she had participated in a recent package tour
to Egypt. Although she often travels to visit family members, it is primarily for
health (and financial) reasons that she has visited Egypt a few times in the winter:
'the climate actually enables me to leave my medication here on the shelves at
home'. Yet these travels have developed into social events too, with Susan acting
as 'personal' guide for friends and her teenage daughter.

The souvenirs in the living room are a mix of things she has bought for herself and things she has received as gifts: ornamented Egyptian ceramic jars and
boxes are displayed on the window sill while an oil lamp from Romania (a gift
from her father), a Hindu God miniature (brought home from her first journey
abroad with her now deceased spouse) and Egyptian perfume and desert sand
co-inhabit a shelf. And souvenirs blend with other symbolic goods on display.
Framed 'old' portrait photographs of her daughter and herself are arranged next to
Greek miniature statues and papyrus (bought at the Egyptian Museum, Cairo) (see

Figure 8.6). These 'souvenirs' (as well as other things) are mementos of people and places she is emotionally involved with. They expose, as Tolia-Kelly (2004b: 685) writes in her home ethnography of migrant women, 'visual cultures that operate beyond the mode of the visual, incorporating embodied memories of past landscapes and relationships'. They are tools for establishing bonds with distanciated significant others; forming part of a 'bricolage', an 'aesthetic . . . that hold[s] considerable meaning, that populates the landscape with clues as to who she is' (Miller, 2006: 237).

In contrast to the front-stage feeling of the living room, her study is somewhat 'cluttered' and 'back-stage-ish'. In addition to containing various 'things', papers and books, the room is dominated by a biggish stationary computer equipped with a webcam. Susan often uses Skype (software that allows free telephone calls over the internet) to contact family members abroad and intermittently distributes odd photographs. There are several traditional photo albums on the shelves. But not all her films have generated photographs – not so far at least: 'I have seven, eight films lying around somewhere just waiting for development . . . [but] I don't think they are of any bloody value. They have probably been lying there for too long by now!' (Susan, home ethnography 9). While desiring photographs to burst into life, she knows that the films are 'almost certainly "dead"'. The films 'haunt' her. While lifeless and useless, they are present and haunting her with guilt rather than the intended 'happy moments'.

New undeveloped films will not haunt her as she turned to digital photography a couple of years ago. Initially, she used her mobile phone and printed the best ones at an electronic equipment store. However, on her recent Egypt holiday, where she travelled with four friends, she used a digital camera. She stores the photographs on a CD-ROM rather than on her computer, 'outsourcing' that initial stage of post-photography work to a travel companion, who collected their cameras and made a CD-ROM for each of them with everyone's photographs: 'one of my friends burns it directly on a disk. He gets all our cameras. And then we can always at one point in time edit them'.

Although the last sentence reveals some loose plans to edit the many hundreds of photographs, a little more than a year after receiving the disk she has still not done any deletion or cropping or given photographs titles or made a folder on her computer. As we noted when she played the beginning of the CD for us, although organized according to day, there were many identical photographs/motifs as well as blurred (or otherwise 'poor') photographs. The photographs are stored but not yet ready for systematic 'retrieval' or attractive 'presentation'. So, although neatly packed and diminutive, this disorganized, great collection will 'haunt' her the day a particular photograph is needed – not unlike the shambolic shoe box.

This lack of editing photo work also explains why the photos have not been distributed, although selected images have been emailed to 'distant ties' and a DVD version of the CD has been made, as Susan prefers to view photographs on her television. However, so far, the only audience for the CD/DVD has been her travel companions, who viewed it at reunion night at her place, when it was shown on the television in the living room accompanied by Arabic food and travel tales.

Next to the computer is the Canon photo printer, with which Susan seems to have an ambivalent relationship. She likes to use it to print her digital photographs, preferring photographs as objects rather than on screens. And, yet, it constantly strikes back at her, mocking her for being lazy and careless: she seldom prints any. Although her Egypt photos are almost a year old now, she admits, a little shamefully, that just one print has been made, so far. Will they be printed in two years' time, we asked her:

Susan:	Oh goodness, at that time, I'll have printed all my photos on paper. I'm certain!
Interviewer:	OK then, do you think that photos are more valuable . . .
Susan:	Yes, because then, you can feel them . . . I think that is nice!

(home ethnography 9)

Here, the respondent claims – in line with Edwards's assertion that the 'aura of thingness' adds value to photographs – that printed digital photographs are more precious when they are printed on high-quality paper. And yet, in practice, she rarely manages to print them. So, not unlike the 'deceased' films, her digital images may also be said to 'haunt' her, because they have not yet reached their intended full potential. However, unlike the films, time is on her side, since digital images do not age – of course her computer might crash or the disk might go missing. Even though she is certain that she will print them in the near future, she might forget or lose interest in them. Only time will tell what afterlife her images will face!

We now move into another home where souvenirs are also much on display, but where the post-photography work is more completed. Within the last decade or so, Jane (home ethnography 8) and her husband, both in their early 60s and retired, have travelled frequently to explore places of cultural and historical significance, often together with a couple they have known for 38 years. Their tourist travel biography has changed from inclusive package tours to Spain and Greece to more independent package tours and 'exotic', faraway destinations more or less off the beaten track such as Romania, South Africa, Peru, Bolivia, Chile, Turkey and Egypt. And their next trip is to China.

In the living room of their neat semi-detached house on the outskirts of a provincial town, handicraft souvenirs and original posters blend stylishly with non-touristic paintings and various decorative objects. A collection of guidebooks sits next to novels on the bookshelf. On moving into the smallish, 'back-stage-ish' study and 'guest room/store room', the presence of photography in their home as well as the materialities of digital photography become manifest. It is in the study that post-photography work is performed and the materials needed for such work are at hand: computer, photo printer, print ink, new CDs, a plastic album with CDs and printed images in plastic folders (Figure 8.7). These are some of the material objects of digital photography.

It is our respondent, the woman of the household, who does all the post-photography work. Although she knows how to delete and edit images, she – like

Figure 8.7 The materialities of digital photography.

the respondent above – does not undertake such 'cleaning' and 'face-lifting' work. Her post-photography work involves picking the best images and organizing them as a PowerPoint presentation[4] (Figure 8.8). Although she occasionally prints some of the best images in colour, the PowerPoint presentation is her preferred way of constructing a coherent, pleasant and 'endurable' presentation of a specific

Figure 8.8 PowerPoint album.

journey, something that she has done for her last many journeys: 'it requires a lot of work . . . but I like it'. She manages the CDs by writing the country and the year on the front with a pen and by storing them in a special plastic CD folder as well as on the computer. The photo printer can also *scan* analogue photographs, and she has made PowerPoint presentations of the pre-digital photography journeys too, thus effectively dematerializing photographic objects, even though the original remains 'material' in the photo album. Such photographs live both material and immaterial lives. Although inhabiting a 'networked computer', her photographs and PowerPoint presentations very rarely travel the internet. Several times, she emphasizes that there is no intended or desired audience for the PowerPoint presentations and prints besides themselves, which also explains the lack of displayed photographs in the front-stage-ish living room. She passes a copy of her presentations only to the friends with whom they travel – in person! One of the few she shares photographs with is a Peruvian woman they met while travelling in Peru seven years ago. They 'Skype' and exchange emails every second month or so. They hope to meet up soon now that her Peruvian friend has started a family in Holland with a Dutch man. In the study, she shows us a plastic framed colour print of her friend's family that she received by email and then made ready for print, storing and display (see Figure 8.7, bottom right).

Contra Edwards (and the respondent above), this respondent does not believe that photographic objects (whether analogue or printed) outshine virtual images. In fact, she prefers her PowerPoint presentations because they reside on her much-used computer; she passes by and visits them more frequently than her old albums on the top shelf in the little-used guest room/store room. Although she occasionally goes through the albums and speaks affectively about a coffee table with transparent glass decorated with old holiday photos, she would not be particularly sad to deposit them, yet 'they would be destroyed if we put them in the loft . . . they cannot take the damp'. Rather than expressing a nostalgic yearning for photographic objects, she complains 'that the passing of time makes them brown, *discoloured*'. For this respondent, time literally destroys the image little by little whereas Edwards would say that such ageing adds to its charm and aura.

In contrast, 'ageless' digital photographs never age. By scanning (digitizing) her old analogue holiday photographs, Jane saves them from further deterioration; photographs arrest time and now scanning can arrest the inescapable wear and tear of photographs. And now that a duplicate dwells in a nicely presented and appreciated PowerPoint presentation, the once so beloved original lives a somewhat forgotten and quiet afterlife; but old love and its fragility save it from humiliating storage in the humid loft. Although still present, the old photographs are somehow absent, seldom seen and noticed. We now make one quick tour into the study of one other family for whom the internet plays a central role in post-photography work.

Anne and Michael (home ethnography 7) are in their early 60s and live in a suburb on the outskirts of a provincial town. Anne works night shifts every other week at a nursing home, whereas Michael retired three years ago. This gives them plenty of time to tour Denmark with their caravan, but they also frequently travel abroad. They visited Alanya in the late 1990s and have been on two package tours to Egypt (to Hurghada and Sharm el Sheikh) within the last couple of years. When we enter their home, coffee and two bulky photo albums (from their first Egyptian trip) wait for us in the conservatory, and Anne is wearing her gold necklace as well as other pieces of jewellery bought in Egypt (Figure 8.9, top). The necklace is one of her favourite and most used pieces of jewellery and it is 'not worn for the occasion [the interview]' (Anne). Once Anne and Michael begin to interact, flicking through the albums, they both get excited about reviving holiday memories and tell travel tales non-stop that far exceed the content and moment of each photograph. The memories and meanings they articulate through, and attach to, the moment fixed by the images exhibit spatial and temporal flexibility. And their tales far surpass the visual; while looking at photographs, they recall the smell of landscapes, the taste of the food, the temperature of the sea, the heat of the sun, and so on.

At one point, Michael recalls buying souvenir t-shirts, and he asks us whether we want to see them. We assent and he walks into the bedroom and returns with a pile of worn t-shirts (two of them from Egypt). One of them is his favourite 'work t-shirt' but he also likes to wear them when they are camping in Denmark because they always initiate 'travel talk' with fellow campers. Both of them state that they do not like purely decorative souvenirs that will only end up 'cluttering' their home; they prefer souvenirs – such as t-shirts and jewellery – that can be used and

Figure 8.9 Album with digital photographs and necklace.

worn as part of their everyday life. As a result, there are no souvenirs on display on the walls or bookshelves in the living room.

Although our 'photo talk' revolves around a traditional photo album, it actually consists of printed digital photographs (Figure 8.9, bottom). They explain that they got them printed in Egypt. While visiting a photo shop to have their photographs burned onto a CD in order to free up space on their camera's memory card, they 'could not resist the temptation to see them on print now – even though it was rather expensive' (Anne). When asked if they prefer physical photographs to seeing them on the computer, they reply:

Michael:	It is much better fun on the computer and TV. Anne has become a bit of nerd. She makes CDs and DVDs and we make it a screen saver on the TV . . . so we can watch it on a huge TV-screen where they fade in and out, with music, too.
Anne:	Yes, that's rather nice . . .
Interviewer:	Interesting! Can I see it?
Anne:	Sure . . . but I have so far only made CDs for Egypt. But we can see the one I have made from Sweden, where we went recently, on the television . . .

(home ethnography 7)

We walk into the living room and Michael puts the CD on: photographs from their Sweden trip fade in and out while a Madonna tune provides the soundtrack. Although disagreeing with the term 'nerd', she explains that she has become very fond of 'playing with digital images' and 'mixing' them with music bites. She probably objects to the term because of its technical connotations; she is interested in 'autobiographical remembering' and 'story-ing' but not in 'image doctoring'. Although never working *upon* images (e.g. cropping or light balance), she does a lot of photo work *with* photographs.

We then walk into their study, which is dominated by a long table, on which are two identical flat-screen computers, and which also contains a photo printer and a pin board covered with printed digital photographs (Figure 8.10). In addition to making DVDs for the television screen, Anne also 'has much fun' with making creative PowerPoint presentations of their holidays, which she emails to interested friends and family members (Figure 8.11). She has, for instance, made one of their recent trip to Egypt in which friends and family members can 'travel along' with them – again with a musical soundscape – on the plane, riding camels, fooling around on a boat, and so on. Anne makes and distributes such DVDs and CDs partly to save and share memories and partly because she thinks such photo work is 'great fun' and 'nice way of being a little creative'. Some of the holiday photographs are also exhibited on their family home page, where they blend with and into their wider family history. And very recently, as we discovered some time after the interview, both of them have set up Facebook profiles, and their photographs have already begun to find a home there, too.

Conclusion

Although studies of tourism often stop when tourists board their return flights, building on our method of mobile tourism ethnographies, this chapter has taken

Figure 8.10 Anne's study.

us further into the everyday lives and domestic spaces of tourists, to explore the *afterlife* of tourism with a particular focus on souvenirs (broadly defined). We have shown that tourism performances are experienced not only for themselves and at one particular moment, but also as a future memory when they become part of the everyday.

We began by highlighting the limitations of traditional accounts of 'trivial souvenirs'. Although many souvenirs undeniably may well be trivial and meaningless, they can equally well be important and precious to their owners, no matter how trivial they appear to the 'outsider'. Yet they are not necessarily meaningful or 'in place' for long. We have seen that souvenirs have mobile cultural biographies as they 'age', travel and take more or less different material and digital forms but also because the social biography of the owner is dynamic. Like the human body, the passing of time marks souvenirs, changes their physical appearance and, by implication, their cultural value: some become more 'valuable' while others are binned or off to the flea market. While some things are given away, disposed of, or used up, other things live a long and attenuated afterlife in the corners of the living room, often without voicing their 'origin' in tourist travels too loudly.

Although souvenirs are objects of the past we do not – or cannot – deal with them here and now. The memories and narratives that souvenirs afford travel in time as people move through life. A given souvenir can bring about many stories and emotions as time passes by and people move through life. We have also shown how the afterlife of tourist photographs is changing dramatically. Whereas tourist

Figure 8.11 Slideshow.

photographs used to be fixed material objects with a secure stable home in the album that as a rule never leaves the house, most are today variable digital objects facing unpredictable afterlives in computer trash bins, folders, email accounts, blogs and, increasingly, social networking sites. What is certain is that tourist photographs have become more visible, mobile and tied up with everyday social-izing on various networked screens.

9 Tourism mobilities and cosmopolitan cultures

Introduction

We began this book by observing the persistent presence of the Other in contemporary consumption and everyday life and noted that the dense webs of corporeal, imaginative and virtual mobilities that weave together geographically distant parts of the globe in some ways have transformed most of us into 'everyday cosmopolitans' (Beck, 2006). Thus, Beck has argued that the increased day-to-day experience of the global through consumption, media, and so on produces a 'banal cosmopolitanism', which, whether it is reflexive or not, 'transforms the experiential spheres of life worlds enforcing a globalization of emotions and empathy' (Beck, 2004: 151–2). In continuation of this, we may argue that tourist performances and experiences are perhaps the single most significant vehicle for the emergence of cosmopolitan orientations in the midst of the everyday. It is when performing tourism that we are most likely to achieve a first-hand experience of other places, people and cultures and experience our connectedness with (or detachment from) these. In this way, tourism is an important part of the ways in which people position themselves in – and as part of – the world.

As we have seen throughout this book, tourism affords different modes of dwelling and of positioning oneself in the world. But do increased leisure travel and tourism also lead to increased awareness of global connectedness and inter-dependencies? Or do they reinforce hegemonic Orientalist scripts about Western superiority by transforming the globe into one big department store for the wealthy and (presumably) Western consumer? This is a concern that has – often tacitly – haunted theories of tourism. In his history of vacation, Orvar Löfgren captures this ambivalence of contemporary mass tourism by claiming that:

> [V]acationing carries an emancipatory potential. At times new forms of mass tourism hold out the hope of changing the world, turning locals into cosmopolitans, breaking down artificial boundaries between nations, localities, classes, or generations – creating global communities. But moving out can also be a way of staying the same. While the skills of becoming a cosmopolitan can be important cultural capital for some, they don't work for others. Tourism can both open and close the mind.
>
> (Löfgren, 1999: 269–70)

Generally speaking, tourist studies has approached tourism as a way of consuming cultural difference and it has focused on the symbolic and discursive aspects of tourism. Being caught up in a representational paradigm, social and cultural tourism theories have largely neglected the mundane, the embodied and the habitual aspects of tourism. Everyday issues of tourist performances have received little attention (see Chapters 2–4). Recent accounts of tourism have sought to dislocate attention from symbolic meanings and discourses to embodied, collaborative and technologized *doings* and *enactments* (see in particular Chapter 4 on this).

The spaces of tourism are not only effects of encounters between 'naked' bodies and material landscapes; they are also inscribed with pre-existing cultural stories, memories, norms, fantasies, family networks, (post)colonial relations and commodity chains, and they are haunted by war and terror as well as bad memories and press (e.g. the so-called cartoon controversy discussed in Chapter 5). These are not necessarily 'present' in place, but they have *effects* in place. They circulate and move and their meanings may change through a variety of local contexts, when they travel across the globe and the global internet as consumer goods, digital photographs or 'global news'. As demonstrated in Chapter 5, the consumption of 'difference' often assumes the mobilization of various forms of 'practical Orientalism' to make 'Other' bodies, objects and spaces manage-*able* and control-*able*. However, 'the Other' may also emerge in our everyday life in unpredictable ways, as when objects, bodies, narratives and images are circulated and connect the 'exotic' setting of tourist performances with the spheres of 'home'.

Rather than transcending the mundane, we have shown that tourism is fashioned by culturally coded escape attempts. Tourists bring not only their own bodies, loved ones and quotidian habits: they also bring the cultural stories in which their everyday lives are immersed. Tourist places, such as the beach, the shisha café, the pool or the famous attraction afford spaces in which a variety of social and cultural stories can be enacted and performed. In Chapters 6 and 7, we followed in the footsteps of tourists and examined how such performative spaces afforded stages onto which a variety of tourist plays and fantasies, of 'the exotic', of intimacy and of everydayness, could be enacted and re-enacted. We also showed how tourists as active performers produce personal photographs to be instantaneously consumed on the screen. Finally, in Chapter 8, we 'followed tourist souvenirs' (including photographs) into and beyond the networked home to explore the symbolical and practical afterlife of tourist objects within the everyday domestic order and, in doing this, 'colonizing' the times and spaces of non-tourist everyday life.

Throughout this book, we have demonstrated how we might think of tourism in terms of networks and flows rather than bounded regions and fixed travel time. Incidents such as the once local cartoon crisis that became global news (see Chapter 5), sending live postcards by MMS messages, uploading photographs to blogs and social networking sites and bringing home souvenirs to decorate shelves, walls and computer screens in houses and flats remind us that tourism is a culture of circulation and connections (see also Chapter 3).

Local tourism performances – at home or away – are framed by and draw upon global flows. As shown in this book, the incorporation of tourist performances, places and things into people's everyday life is a significant element in 'doing tourism'. Touristic social life centres around everyday routines and habitual dispositions, and tourist things and images migrate into living rooms and studies where they populate, more or less permanently, and more or less pleasurably, our bodies, living spaces and networked screens.

In this short final chapter, elaborating on the previous chapter and drawing on our 18 home ethnographies, we bring out how people mobilize tourist performances to situate themselves in the world as part of cosmopolitan cultures. In what follows, we show how people use the afterlife of their holiday to position themselves and their everyday life within cultural geographies of travel and tourism.

Routing cosmopolitanism

As Beck (2002, 2004, 2006) points out, the novelty of contemporary cosmopolitan elements in everyday life is that 'issues of global concern are [now] becoming part of the everyday local experiences and the "moral life-worlds" of the people' (Beck, 2002: 17). Following on from this, Beck argues that research ought to engage with the actually (empirically) existing cosmopolitanism in its differentiated, situated, materialized and embodied forms. Beck's suggestion parallels developments also found in other forms of social and cultural studies (see, for example, Appiah, 1998; B. Robins, 1998; Skrbis *et al.*, 2004; Molz, 2006b). Instead of generalizing utopian ideas about cosmopolitanism, such studies suggest that it might be better to imagine cosmopolitanism in a plural sense: as situated and particular embodied habits of thoughts, feelings and dispositions towards the world. For example, Appiah suggests the idea of 'rooted cosmopolitanism', which, he argues, entails:

> The possibility of a world in which *everyone* is a rooted cosmopolitan, attached to a home of his or her own, with its own particularities, but taking pleasure from the presence of other, different, places that are home to other, different, people.
>
> (Appiah, 1998: 91)

Appiah's notion of 'rooted cosmopolitanism' points to the possibility of belonging within a geography of multiple, networked places. Paraphrasing Clifford's famous play on words (see Chapter 1), we may take this idea one step further by suggesting the notion of '*rou*ted cosmopolitanism' as a way of grasping the materializations and mobilizations of cosmopolitanism in people's everyday lives. What we suggest is to trace out the various empirically and contingently produced 'cultures of cosmopolitanism' that are currently emerging in everyday life and consumption. By 'routing cosmopolitanism', we hope to be able to address the various *paradoxes* of cosmopolitanism.

In doing this, we have to leave behind idealistic notions of cosmopolitanism and empirically 'flesh out' the many ways in which people position themselves in a still more connected, networked world of mundane mobilities and flows. This points to a form of cosmopolitanism that is empirically diverse and takes many particular forms. Cosmopolitanism cultures are afforded by the particular routes people travel, the social and spatial connections and networks they make and the flows of images, objects and consumer gods that feed into their everyday homes.

Drawing on our home ethnographies of 'ordinary tourists', we will distinguish between three forms of cosmopolitanism in this chapter. In addition to the *aesthetic cosmopolitanism* emerging from the globalization of consumer preferences and tastes as well as the democratization of travel (Lash and Urry, 1994: 256–7; Nava, 2002), we identify two other 'cultures of travel'. *Orientalist cosmopolitanism*, we argue, enables people to enjoy (and claim) travel and tourism as a right and at the same time enact elements of 'practical Orientalism' in their tourist performances away and at home. *Connective cosmopolitanism* similarly transgresses the idea of *aesthetic cosmopolitanism* as tied up with consumption. The notion of *connective cosmopolitanism* points to the capability of (some) people to 'root' (cf. Appiah above) and belong in multiple places simultaneously.

Aesthetic cosmopolitanism

Traditionally, this notion of 'cosmopolitanism' connotes elitist utopias of friction-less travel and communication and global community (see also Cresswell, 2006: 222). In their research on 'cosmopolitan cultures' Szerszynski and Urry (2006) suggest that cosmopolitan practices and dispositions consist of a set of characteristics which, in addition to extensive mobilities, include curiosity about places and ability to consume and compare places; an openness towards 'other' people and cultures; and practices and dispositions that enable at least certain groups of people to 'inhabit the world from afar'. The language of such cosmopolitanism, they emphasize, is 'a language of mobility, of abstract characteristics and comparison' (ibid.: 127).

Such relations between cosmopolitanism and tourist travel are clearly visible in some of the homes we visited, as described in Chapter 8. They are materialized in living rooms by objects and souvenirs on display, such as wall world maps, guidebooks and travel literature, globes, water-pipes, cushions and rugs (Figure 9.1) and verbalized in their often knowledgeable travel and souvenir accounts (see Chapter 8). Being part of – and being able to navigate within – a 'global culture' is central to such respondents.

In the home of Keith and Lisa, a young couple living in Copenhagen, such a cosmopolitan culture is materialized by an abundance of 'souvenirs with a use-value' (see Chapter 8). This is not the case in Ann's home. Apart from some travel literature on the shelves, her living room is dominated by a huge world map, used for pointing out travels of friends of the family to her daughter Pennie and 'of course to see Africa, with Egypt [which they have just visited] as the reference point' (home ethnography 2). To these respondents, travelling, and being part of

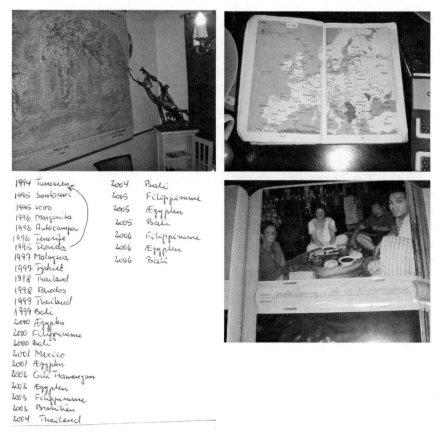

Figure 9.1 Positioning.

a travelling culture, is materialized not only in the purchase and use of personal, useful and meaningful souvenirs, but also by, for example, pinpointing places visited (Figure 9.1, upper left corner). Similarly, Keith and Lisa draw up the routes of their first travel together in a guidebook, hence inscribing their relationship as a couple on the guidebook map over Turkey (Figure 9.1, top right).

People 'travel together' to mark important and special moments in their lives. For Keith and Lisa, the trip to Turkey was their first travel together after having met each other and, for Lisa especially, the trip was also a 'test' whether she would be able to cope with travelling to non-Western countries. To Ann, her trip to Egypt with one of her daughters was explicitly intended as an inauguration rite marking the first steps of her daughter into adolescence, an intention she expresses with the formula: 'when one is about to begin school, one gets to travel alone with her mother' (home ethnography 2). Having previously travelled extensively in Europe, Egypt, Jordan and Israel, she explains that this is a good opportunity for her daughter Pennie to gain first-hand knowledge of 'other' places:

> The idea was something like: I had been terribly busy writing my masters report . . . and then this seemed like an idea that could be a way of getting in tune with each other's lives again . . . before she would start school . . . which is . . . something big, right . . .
>
> (home ethnography 2)

The choice of marking the transition of her daughter from child to schoolchild by travelling was not accidental. The decision to travel to Egypt rested on an extensive comparison of what this destination 'offers' in relation to other destinations. In this extract, she compares it with Paris:

> Well, I thought a lot about Pennie and Egypt . . . the desert, camels, that kind of thing . . . I thought this would be something that would be very different from . . . Paris was mentioned too, but I had it like, well, no, we can do that another time when she is a teenager, then she can bring a friend, and leave me alone. Because . . . I simply do not think Paris is able at . . . leaving an impression [for a child] . . . Whereas, I thought, mummies are potent, the big temples, the desert, camels and the smells. And like this, there are many things inviting that you can construct . . . mark that this is an extraordinary experience you have had . . .
>
> (home ethnography 2)

Although the trip to Egypt was explicitly performed to bond with her daughter, the destination was carefully chosen on the background of the affordances offered to a child taking her first steps into the culture of travel. Thus, 'inhabiting the world from afar' (see Szeszynski and Urry, 2006, above), verbalizing and visibly displaying intimate knowledge and capabilities of comparing and navigating 'distant' places are central to Ann and Pennie (as to Keith and Lisa discussed earlier). Thus, they are illustrative of *aesthetic cosmopolitanism.*

As more and more groups in society have gained access to cheap air travel around the globe, the ability to experience, compare and discriminate between places and cultures is not reserved for the educated elites. In many homes, we find traces of such an aesthetic cosmopolitanism in the ways that people ornament their homes and emphasize the rights (and capability) of travelling. Yet, often such expressions are articulated in more ambiguous terms, as we will now demonstrate with the notion of Orientalist cosmopolitanism.

Orientalist cosmopolitanism

As demonstrated in Chapters 2 and 6, tourist travel often includes the enactment and performance of 'banal' stereotypes of Self and Other. Furthermore, as shown in Chapter 5, Orientalist tropes and techniques have been decisive in tourism for transforming other places, people and cultures into consume-*able* objects. In many of the homes we visited, we encountered a strange and somewhat paradoxical coexistence of 'cosmopolitan' and 'Orientalist' statements. While claiming the right to travel and tourism, these respondents would – more or less reflexively – enact elements of 'practical Orientalism' in their tourist performances away and at home.

Speaking about their trip to Egypt, Leigh and Ann-Lee, an elderly couple living in a small town outside Copenhagen (see Chapter 8), explain that they had to stay in the resort area or on their cruise ship during their trip to Egypt, because beyond these patrolled spaces, danger lurked:

> I always made sure to be together with some other people. I would say it is really frightening. You cannot move anywhere in Egypt without passing road barriers and barb wire.
>
> (home ethnography 6)

For Leigh and Ann-Lee, the insecurity of walking the streets of Egyptian cities is mainly interpreted as a consequence of Egypt's relative disconnection from the global economy. However, when discussing their holidays in Turkey, they elaborate on this discomfort by linking the places with the people populating them:

> [there is] a big difference between being in Turkey and Egypt. In Turkey, the people who have been living up here [Denmark] for some time, they are really disgusting, when they are in their homeland. Unpleasant like hell.
>
> (home ethnography 6)

Leigh and Ann-Lee's travel accounts are in many ways emblematic of Bauman's (1998) notion of the world as 'the tourist's oyster'. Their trips to the different corners of the world rely on sharply dichotomized divisions between safe and unsafe spaces and 'us' and 'them', mobilizing the kind of 'practical Orientalism' discussed in Chapter 5. This 'Manichean Geography' (Gregory, 2004) was also expressed by other respondents, who similarly explained it with reference to their encounters with migrants 'at home' in Denmark. For example, Susi, a woman in her 30s from Aarhus, similarly utters her uneasiness with people in Egypt and Turkey by relating an incident in which she and her mother had been assaulted in an Egyptian street to her encounters with 'other bodies' in Denmark:

> I know how it is in my youth club [her work place]: when there is an arrangement of some sort, then they [the Turks] turn up in groups from Gjellerup Parken [a renowned ghetto area in Aarhus, Denmark]. And they do not have any respect for anything . . . especially not the way they approach women.
>
> (home ethnography 8)

As a way of positioning oneself in the world, this *orientalist cosmopolitanism* was most clearly expressed when we visited Heinz, a retired worker and widower, living on his own. Since his first tour outside Europe in 1994, just after his younger son moved out, he has been a frequent traveller. Before our visit in November 2006, he had already been to the Philippines, Egypt and Bali that year and he was busy packing to go to Thailand the day after our visit. Heinz keeps an archive of systematic reports of the places he visits. Every evening on his travels, he fills one sheet of paper with his impressions from that day. The sheets are ordered into the same blue folders, and kept in the cupboard beside the television together with tickets, photocopies of travel brochures from tour operators, and so on. He also documents his holiday on video tape. He explains this systematic recording of his travels as partly 'because I like to compare what they write and how things are' and partly because 'it is becoming difficult to distinguish between, different sorts of things, remember how things were in a specific country in, say, 2004' (home ethnography 13).

For Heinz, like the other people discussed above, the ability to organize travel, distinguish and evaluate places, cultures and people as well as to appreciate differences in global society are all high-ranking skills and values. However, places to travel to are assigned very specific roles in the cultural geography within which he positions himself. This is also mirrored in the repetitive patterns of his travel career: Thailand is for relaxing, Bali for exoticism and Egypt for diving and so on (Figure 9.1, bottom left). The roles Heinz assigns to places are firmly rooted in the embodied, sensual and habitual performances of what we have called 'practical Orientalism'. This is clearly seen in his distinction between Egypt (where he only goes for diving trips) and Bali ('an inviting island in a sea of Muslims'). He talks passionately about his intimate and lasting relationship with a Balinese family, while at the same time explaining:

> No, I do not talk to people when I am in Egypt. You know I have only met the diving instructors and they are young men, and our everyday lives would be too remote from each other to have anything to talk about.
>
> (home ethnography 13)

Heinz does not explicitly relate to experiences of multiculturalism and migrants in Western Europe, but will touch upon it occasionally when we are talking about other issues. For example, when discussing the cartoon controversy (see Chapter 5), he argues: 'Well, I do not want to accuse ALL Muslims, the diver instructors for example are absolutely fine, but . . .'. Or talking about debates on multiculturalism in Denmark: 'Well, I do not want to project the debate we have in Denmark onto the people living there, but it may be something that sticks to you . . .' (home ethnography 6).

Although Heinz admits that his view on migrants in Denmark may be 'something that sticks to you' when travelling, other people emphasize how the habitual routines of practical Orientalism had been challenged and transformed by the memories and experiences brought home from their tourist travels. This was most clearly expressed by John and Sara. Both had lived in so-called 'ghetto areas' in Denmark – urban areas stigmatized for their high proportion of Middle Eastern and Third World immigrants – and worked as factory employees together with people of multiple ethnic origins. When talking about how tourist experiences made on their frequent holidays in Turkey related to broader aspects of their everyday lives, Sara and John explain:

Sara: But don't you remember, John, in the beginning . . . the first time we were going to Turkey . . . we were walking through the shopping centre and there . . . with all these Turks hanging around everywhere . . . and you said: 'what the hell are we going to Turkey for. We've got plenty of these people here [in Denmark]'.

John: Yeah, well, it is all those people they do not want to keep down there [Turkey] that they are sending to us up here [Denmark] . . . And then when you actually get down there it is plainly speaking like getting a bucket of cold water thrown in your face: 'Well, what is it actually you've been going around saying?'

(home ethnography 15)

As seen in these examples, the democratization of mass travel and tourism does carry with it a paradoxical and even sometimes cynical element, as travelling and consuming places become verbalized as a right for some while 'others' are stereotyped as servants, even annoyances, for the tourist. By suggesting the notion of *Orientalist cosmopolitanism*, we attempt to capture this 'dark side' of cosmopolitanism. In the next subsection, we briefly outline the third and final particularization of cosmopolitan cultures encountered in the lives of our respondents.

Connective cosmopolitanism

As we have seen above, the notion of 'cosmopolitan culture' contains ambiguities and paradoxes. Moreover, cosmopolitan elements emerge both in relation to (elitist) global consumer cultures and in (paradoxical) coexistence with Orientalist scripts and habitual routines. We have also outlined how some of our respondents reflected on the encounter between the 'practical Orientalism' brought with them on travel and the encounters with the Other brought home and translated into the everyday spaces 'at home'. To John and Sara, discussed above, their holidays in Turkey have in important ways changed their way of relating to and encountering people of 'other' cultural backgrounds. Thus, the practical Orientalism that shaped their relations to immigrants earlier has been superseded by what could be called a *connective cosmopolitanism* based on situated experiences and practices of interaction with particular 'others'. The notion of connected cosmopolitanism has affinities with Appiah's notion of a 'rooted cosmopolitanism', entailing the possibility of inhabiting and dwelling in multiple places. Thus, the notion of 'connective cosmopolitanism' may force us to think of cosmopolitanism as something that relies on and reproduces a sustained commitment to particular places and people instead of taking pleasure from being part of an abstracted global or cosmopolitan culture.

This also differentiates this type of positioning oneself in the world from the highly *positive* valuation of 'otherness' enacted through 'aesthetic cosmopolitanism' and *negative* ones found in the ambivalent culture of 'Orientalist cosmopolitanism'. Although they are very different, what 'aesthetic cosmopolitanism' and Orientalist cosmopolitanism have in common is that they both draw on (respectively a pleasurable or repulsive) *distance* between well-known/exotic, tourism/everyday life, tourists/locals, Self/others, whereas *connective cosmopolitanism* is based on the sustained *connection* with other people, places and cultures, and the incorporation of these into the fabrics and routines of the everyday at home.

In Chapter 6, we saw how some tourists build up their memories of, and social relations with, tourist places over a sustained period of time, transforming these into 'homely' places of belonging and care. By returning to a particular place year after year, making friends with locals (or other repeat tourists) or owning/pursuing second homes, people moor parts of their everyday life and concerns to particular tourist places. Furthermore, in Chapter 8, we saw how the connections of tourism may enter into people's everyday life through Skyping, emails and text messages, and symbolical decorations on walls, screens and bodies (see especially the discussion of Susan's living room; home ethnography 9, Chapter 8). This enables people to connect more directly to other people's everyday lives and concerns. For example, Susan explains that on her repeat visits to Egypt she prefers to spend time at 'local' cafés discussing everyday issues with people she meets and some of these discussions may continue over the internet, whereas at home:

> I prefer to sit at the local coffee shop, have a pipe of tobacco, and a cup of Arabian coffee . . . I mean . . . that's much more cosy . . . [we talk], for example, about divorces . . . when I discuss with other women . . . Now take these cartoons as an example, they tend to laugh a bit, but I guess that they are not the worst kind of . . . people working in the resort areas . . . they are much more engaged in supporting their families . . . So we discuss divorces, politics and so on.
>
> (home ethnography 9)

The culture of *connective cosmopolitanism* was expressed most clearly by Hanna. In the last couple of years, Hanna has been travelling to Egypt regularly in connection with the purchase of a holiday house in Hurghada on the Egyptian Red Sea shore. As her husband has been too busy with their business in Denmark, she has had to supervise the construction work on her own. At the same time, her visits have facilitated friendships with people there, and in Cairo friends occasionally invite her to family events and on visits to their own families in suburban Luxor, Cairo and Alexandria. Speaking about how she and her two children (with whom she travels) have got in contact with the 'friends' that populate her photo album (see Figure 9.1, bottom right), she explains:

> Well, they are very eager to seek contact, you know, on the street, so some of them we just met, and if I then feel that they seem alright, I do not have any second-thoughts about letting them suggest their ideas . . . And I am not

worried about . . . that, you know they like to invite you to their home, and show what they've got . . . so it's simply to follow . . .

(home ethnography 18)

She has regular telephone and email contact with her friends (two in Cairo, two in Hurghada) every fortnight, mostly about everyday issues such as work and the progress of the house, but they also 'talk' at the time of important moments such as national holidays (her Egyptian friends are very eager to remember Christmas and Easter holidays) or when incidents such as the cartoon controversy or bombings (Sharm, Dahab) hit the global media circuit. Apart from that, communication revolves around mundane issues such as their children, how business is going, and so on:

We do have sustained contact . . . I should actually have been to Egypt in February when all flights suddenly were cancelled [as a consequence of the 'cartoon controversy']. And how I felt about it . . . Well. I was very curious about what they thought about it all, in Cairo. We were in daily contact . . . and I was thinking about what he was thinking about it, so I did send him a text message, just to hear what the guy we have most contact to thought about . . . and I got a very fast reply saying that 'we are fully aware that not all Danes are like that' . . . So no, it is very reasonable . . . of course some have been critical . . . we were there just afterwards, but no one commented on it to us . . . When we asked . . . they were fully up to date about the discussion, but no . . .

(home ethnography 18)

Conclusion: tourism mobilities and cosmopolitan cultures

Throughout this book, we have shown how tourist performances in significant ways feed into the practices and relations of people's everyday lives connecting spaces of tourism and spaces of home through a variety of corporeal, virtual and material mobilities. Tourist performances are not limited to the two-week holiday of the pleasure-seeking traveller, but are integrated into wider aspects of consumption and everyday life.

In this chapter, we sought to bring out the effects of tourism mobilities on the way people position themselves within cultural geographies of travel and tourism, and how tourism may be said to induce cosmopolitanism aspects on people's everyday lives. As we have seen, the notion of 'culture of cosmopolitanism' is an ambivalent one containing contradictions and paradoxes that may both erode and reinforce stereotypical imaginative geographies.

The cultures of cosmopolitanism enacted and the households discussed in this chapter and the way these 'route' the mobile connections between tourist places 'away' and the everyday 'at home' are very different. In addition to the *aesthetic cosmopolitanism* which has often been noted to be an integral part of the rise of modern tourism, we have suggested the notions of *orientalist cosmopolitanism* and *connective cosmopolitanism* to capture the different ways in which cosmopolitanism emerges as part of the everyday in a world in which the 'global' is increasingly part of mundane experiences. Both the notions of Orientalist cosmopolitanism and connective cosmopolitanism point to more non-reflexive and embodied responses to the penetration of everyday life by global mobilities than what is usually addressed in discussions of cosmopolitan cultures. By discriminating between such cultures of cosmopolitanism, we have traced out the real existing cultures of cosmopolitanism (Beck, 2006) enabled by current transformations of everyday life and travel. Thus, 'everyday cosmopolitanism' may produce borders and connections that are drawn and established, in habitual, non-reflexive and embodied ways.

Distinguishing between such differences in the way people respond to and enact the presence of 'the global' as part of the everyday is necessary to avoid the hyperbolic claims that often haunt discussions of globalization, tourism and cosmopolitanism. Tourism does indeed carry an emancipatory potential (as well as its opposites), as we quoted Löfgren (1999) at the beginning of this chapter. The way this potential is embodied, particularized and materialized is, however, an open game.

Notes

2 De-exoticizing tourist travel

1 A portable media player designed and marketed by Apple Inc.
2 A handheld video game console developed and manufactured by Nintendo.

3 Following flows

1 http://www.dst.dk/Statistik/Nyt/Emneopdelt.aspx?si=5&msi=6.
2 http://politiken.dk/tjek/digitalt/internet/article606196.ece.
3 http://en.wikipedia.org/wiki/List_of_social_networking_websites.

5 Mobilizing the Orient

1 http://en.wikipedia.org/wiki/Jyllands-Posten_Muhammad_cartoons_controversy.
2 http://www.trekearth.com/gallery/Middle_East/Turkey/Marmara/Istanbul/Istanbul/
photo945723.htm.
3 http://www.trekearth.com/gallery/Africa/Egypt/photo957717.htm.
4 http://search.fotki.com/?q=giza.

7 Performing digital photography

1 http://www.kodak.com/eknec/PageQuerier.jhtml?pq-path=6433&pq-locale=en_
US&_requestid=6976.
2 http://www.computerworld.dk/art/30029?a=kw_fp&i=24.
3 http://www.computerworld.dk/art/30029?a=kw_fp&i=24.
4 http://nseries.com/nseries/index.html#l=experiences,photo,share_on_ovi.
5 http://www.text.it/mediacentre/press_release_list.cfm?thePublicationID=2C4F
B155-15C5-F4C0-99FCB4EAAED7A798.
6 http://www.text.it/mediacentre/press_release_list.cfm?thePublicationID=2C4F
B155-15C5-F4C0-99FCB4EAAED7A798.

8 The afterlife of tourism

1 http://en.wikipedia.org/wiki/Flickr.
2 http://blog.flickr.net/en/2008/11/03/3-billion/.
3 http://www.facebook.com/press/info.php?statistics.
4 A presentation program developed by Microsoft that is part of the Microsoft Office
suite.

References

Allen, J. (2000) 'On George Simmel: proximity, distance and movement', in M. Crang and N. Thrift (eds) *Thinking Space*, London: Routledge.

Ammitzbøll, P. and Vidino, L. (2007) 'After the Danish cartoon controversy', *Middle East Quarterly*, 14(1): 3–11.

Andrews, H. (2005) 'Feeling at home: embodying Britishness in a Spanish charter tourist resort', *Tourist Studies*, 5(3): 247–66.

Andrews, M. (1989) *The Search for the Picturesque: Landscape, Aesthetics and Tourism in Britain, 1760–1800*, Aldershot: Scholar Press.

Appadurai, A. (ed.) (1986) *The Social Life of Things: Commodities in Cultural Perspective*, Cambridge: Cambridge University Press.

—— (1988) 'Putting hierarchy in its place', *Cultural Anthropology*, 3(1): 36–49.

—— (1999) *Modernity at Large: Cultural Dimensions of Globalization*, Minneapolis: University of Minnesota Press.

Appiah, K. A. (1998) 'Cosmopolitan patriots', in P. Cheah and B. Robbins (eds) *Cosmopolis: Thinking and Feeling beyond the Nation*, Minneapolis: University of Minnesota.

Ateljevic, I. and Doorne, S. (2003) 'Culture, economy and tourism commodities: social relations of production and consumption', *Tourist Studies*, 3(2): 123–41.

Axhausen, K. W., Zimmermann, A., Schönfelder, S., Rindsfüser, G. and Haupt, T. (2002) 'Observing the rhythms of daily life: a six-week travel diary', *Transportation*, 29(3): 95–124.

Bærenholdt, J. and Haldrup, M. (2004) 'On the track of the Vikings', in M. Sheller and J. Urry (eds) *Tourism Mobilities: Places to Play, Places in Play*, London: Routledge.

Bærenholdt, J., Haldrup, M., Larsen, J. and Urry, J. (2004) *Performing Tourist Places*, Aldershot: Ashgate.

Bærenholdt, J., Haldrup, M. and Larsen, J. (2008) 'Performing cultural attractions', in J. Sundbo and P. Darmer (eds) *Creating Experiences in the Experience Economy*, Cheltenham: Edward Elgar.

Ball, M. and Smith, G. (2001) 'Technologies of realism? Ethnographic uses of photography and film', in P. A. Atkinson, S. Dalamont, A. J. Coffey and J. Lofland (eds) *Handbook of Ethnography*, London: Sage.

Barthes, R. (2000) *Camera Lucida*, London: Vintage.

Battarbee, K. and Koskinen, I. (2005) 'Co-experience: user experience as interaction', *CoDesign*, 1(1): 5–18.

Bauman, Z. (1998) *Globalization*, Cambridge: Polity Press.

—— (2000) *Liquid Modernity*, Cambridge: Polity Press.

Beck, U. (2002) 'The cosmopolitan society and its enemies', *Theory, Culture and Society*, 19(1–2): 17–40.

—— (2004) 'Cosmopolitical realism: on the distinction between cosmopolitanism in philosophy and the social sciences', *Global Networks*, 4(2): 131–56.

—— (2006) *Cosmopolitan Vision*, Cambridge: Polity Press.

Beer, D. and Burrows, R. (2007) 'Sociology and, of and in Web 2.0: some initial considerations', *Sociological Research Online*, 12(5), http://www.socresonline.org. uk/12/5/17.html.

Behdad, A. (1994) *Belated Travelers: Orientalism in the Age of Colonial Dissolution*, Cork: Cork University Press.

Bell, C. and Lyall, J. (2002) 'The accelerated sublime: thrill-seeking adventure heroes in the commodified landscape', in S. Coleman and M. Crang (eds) *Tourism: Between Place and Performance*, Oxford: Berghahn Books.

—— (2005) ' "I was here": pixilated evidence', in D. Crouch, R. Jackson, and F. Thompson (eds) *The Media and the Tourist Imagination: Converging Cultures*, London: Routledge.

Berger, J. (1984) *And Our Faces, My Heart, Brief as Photos*, New York: Pantheon Books.

Bhabha, H. (1994) *The Location of Culture*, London: Routledge.

Bial, H. (ed.) (2004) *The Performance Studies Reader*, London: Routledge.

Billig, M. (1997) *Banal Nationalism*, London: Sage.

Bingham, N. (1996) 'Object-ions: from technical determinism towards geographies of relations', *Environment and Planning D*, 14(6): 635–57.

Bleich, E. (2006) 'On democratic integration and free speech: response to Tariq Moddod and Randall Hansen', *International Migration*, 44(5): 17–22.

Blunt, A. (2004) 'Cultural geographies: cultural geographies of home', *Progress in Human Geography*, 29(4): 505–15.

Blunt, A. and Varley, A. (2004) 'Geographies of home', *Cultural Geographies*, 11: 3–6.

Boden, D. and Molotch, H. (1994) 'The compulsion to proximity', in R. Friedland and D. Boden (eds) *Nowhere, Space, Time and Modernity*, Berkeley: University of California.

Bonde, B. N. (2007) 'How 12 cartoons of the Prophet Mohammed were brought to trigger an international conflict', *Nordicom Reviews*, 28(1): 33–48.

Bourdieu, P. (1990) *Photography: A Middle-brow Art*, London: Polity Press.

Brown, B. (2007) 'Working the problems of tourism', *Annals of Tourism Research*, 34(2): 364–83.

Brown, D. (1996) 'Genuine fakes', in T. Selwyn (eds) *The Tourist Image*, Chichester: John Wiley & Sons.

Bruner, E. (2005) *Culture on Tour: Ethnographies of Travel*, Chicago: University of Illinois Press.

Bryce, D. (2007) 'Repacking Orientalism: discourses on Egypt and Turkey in British outbound tourism', *Tourist Studies*, 7(2): 165–91.

Burns, P. M. (2004) 'Six postcards from Arabia: a visual discourse of colonial travels in the Orient', *Tourist Studies*, 4(3): 255–75.

Butler, J. (1993) *Bodies that Matter: On the Discursive Limits of Sex*, London: Routledge.

Caletrio, J. (2003) 'A ravaging Mediterranean passion: tourism and environmental change in Europe's playground', unpublished PhD thesis, Lancaster University.

Callon, M. and Law, J. (2004) 'Guest editorial', *Environment and Planning D*, 22: 3–11.

Cass, J. (2004) 'Egypt on steroids: Luxor Las Vegas and postmodern Orientalism', in D. M. Lasansky and B. McLaren (eds) *Architecture and Tourism. Perception, Performance and Place*, London: Berg.

Castells, M. (1996) *The Rise of the Network Society*, London: Blackwell.

—— (2000) 'Materials for an explanatory theory of the network society', *British Journal of Sociology*, 51: 5–24.

—— (2004) 'Informationalism, networks, and the network society: a theoretical blueprint', in M. Castells (ed.) *The Network Society*, Cheltenham: Edward Elgar.

Castells, M., Fernandez-Ardevol, M., Linchuan, Q.-J. and Sey, A. (2007) *Mobile Communication and Society: A Global Perspective*, Cambridge, MA: MIT Press.

Chalfen, R. (1987) *Snapshot Versions of Life*, Bowling Green, OH: Bowling Green State University Popular Press.

Clifford, J. (1986) 'Introduction: partial truths', in J. Clifford and G. Marcus (eds) *Writing Culture: The Poetics and Politics of Ethnography*, Berkeley: University of California Press.

—— (1997) *Routes*, Cambridge, MA: Harvard University Press.

Cloke, P. and Perkins, H. C. (1998) 'Cracking the canyon with the awesome foursome: presentations of adventure tourism in New Zealand', *Environment and Planning D*, 16(3): 185–218.

Cohen, E. (1972) 'Toward a sociology of international tourism', *Social Research*, 39(1): 163–82.

—— (1979) 'A phenomenology of tourist experience', *Sociology*, 13(2): 179–202.

Cohen, R. K. (2005) 'What does the photoblog want?', *Media, Culture and Society*, 27(6): 883–901.

Coleman, S. and Crang, M. (eds) (2002) *Tourism: Between Place and Performance*, Oxford: Berghahn Books.

Coles, T. and Timothy, D. (eds) (2004) *Tourism, Diasporas and Space*, London: Routledge.

Coles, T., Hall, C. M. and Duval, D. (2005) 'Mobilising tourism: a post-disciplinary critique', *Tourism Recreation Research*, 30(3): 31–41.

Cook, I. *et al.* (2004) 'Follow the thing: Papaya', *Antipode*, 36(4): 642–64.

Cook, I. and Harrison, M. (2007) 'Follow the thing', *Space and Culture*, 10(1): 40–63.

Cooper, G., Green, N., Murtagh, M. G. and Harper, R. (2002) 'Mobile society? Technology, distance and presence', in S. Woolgar (ed.) *Virtual Society? Technology, Cyberbole, Reality*, Oxford: Oxford University Press.

Cosgrove, D. (1984) *Social Formation and Symbolic Landscape*, Madison: University of Wisconsin Press.

—— (2003) 'Landscape and the European sense of sight – eyeing nature', in K. Anderson, M. Domosh and S. Pile (eds) *Handbook of Cultural Geography*, London: Sage.

Couldry, N. (2004) 'Theorising media as practice', *Social Semiotics*, 14(2): 115–32.

Cox, A. M., Clough, P. D. and Marlow, J. (2008) 'Flickr: a first look at user behaviour in the context of photography as serious leisure', *Information Research*, 13(1): paper 336 (http://informationr.net/ir/13–1/paper336.html).

Crang, M. (1999) 'Knowing, tourism and practices of vision', in D. Crouch (ed.) *Leisure/ Tourism Geographies: Practices and Geographical Knowledge*, London: Routledge.

—— (2002) 'Qualitative methods: the new orthodoxy?', *Progress in Human Geography*, 26(5): 647–55.

—— (2003) 'Placing Jane Austen, displacing England: between book, history and nation', in S. Pucci and J. Thompson (eds) *Jane Austen and Co.: Remaking the Past in Contemporary Culture*, New York: SUNY Press.

—— (2005) 'Qualitative methods (part 3): there is nothing outside the text?', *Progress in Human Geography*, 29(2): 225–33.

—— (2006) 'Circulation and emplacement: the hollowed out performance of tourism', in C. Minca and T. Oakes (eds) *Travels in Paradox: Remapping Tourism*, Lanham, MD: Rowman & Littlefield.

Crang, P. (1994) 'Its showtime: on the workplace geographies of display in a restaurant in South East England', *Environment and Planning D*, 12: 675–704.

—— (1997) 'Performing the tourist product', in C. Rojek and J. Urry (eds) *Touring Cultures: Transformations of Travel and Theory*, London: Routledge.

Crang, P., Dwyer, C. and Jackson, P. (2003) 'Transnationalism and the spaces of commodity culture', *Progress in Human Geography*, 27(4): 438–56.

Cresswell, T. (2003) 'Landscape and the obliteration of practice', in K. Anderson, M. Domosh and S. Pile (eds) *Handbook of Cultural Geography*, London: Sage.

—— (2006) *On the Move: Mobility in the Modern Western World*, London: Routledge.

Crouch, D. (1999) 'Introduction: encounters in leisure/tourism', in D. Crouch (ed.) *Leisure/Tourism Geographies: Practices and Geographical Knowledge*, London: Routledge.

—— (2002) 'Surrounded by place, embodied encounters', in S. Coleman and M. Crang (eds) *Tourism: Between Place and Performance*, Oxford: Berghahn Books.

—— (2003a) 'Spacing, performing, and becoming: tangles in the mundane', *Environment and Planning A*, 35(11): 1945–60.

—— (2003b) 'Performances and constitutions of natures: a consideration of the performance of "lay geographies"', in B. Szerszynski, W. Heim and C. Waterton (eds) *Nature Performed: Environment, Culture and Performance*, London: Blackwell.

Crouch, D., Aronsson, L. and Wahlström, L. (2001) 'Tourist encounters', *Tourist Studies*, 1(3): 253–70.

Cwerner, S. B. and Metcalfe, A. (2003) 'Storage and clutter: discourses and practices of order in the domestic world', *Journal of Design History*, 16(3): 229–39.

Dann, G. (1996a) 'The people of tourist brochures', in T. Selwyn (ed.) *The Tourist Image*, Chichester: John Wiley & Sons.

—— (1996b) *The Language of Tourism*, Wallingford: CAB International.

Dant, T. (1998) 'Playing with things: objects and subjects in windsurfing', *Journal of Material Culture*, 3(1): 77–95.

—— (1999) *Material Culture in the Social World*, Buckingham: Open University Press.

—— (2004) 'The driver–car', *Theory, Culture and Society*, 21(4–5): 61–79.

—— (2005) *Materiality and Society*, Maidenhead: Open University Press.

Dateline (2003) *Design and Application of a Travel Survey for European Long-Distance Trips Based on an International Network of Expertise*, Munich: Institut für Verkehrs- und Infrastrukturforschung (available at http://www.ncl.ac.uk/dateline/home_page.htm).

Davies, J. (2006) 'Affinities and beyond! Developing ways of seeing in online spaces', *E–Learning*, 3(2): 217–34.

de Botton, A. (2002) *The Art of Travel*, New York: Pantheon Books.

de Certeau, M. (1984) *The Practice of Everyday Life*, Berkeley: University of California Press.

della Dora, V. (2007) 'Putting the world into a box: a geography of nineteenth-century "travelling landscapes"', *Geografiska Annaler*, 89B(4): 287–306.

Denzin, N. K. (1997) *Interpretive Ethnography: Ethnographic Practices for the Twenty-first Century*, London: Sage.

Denzin, N. K. and Lincoln, Y. (2003) 'Introduction: the discipline and practice of qualitative research', in N. K. Norman and Y. Lincoln (eds) *The Landscape of Qualitative Research*, London: Sage.

Desmond, J. C. (1999) *Staging Tourism: Bodies on Display from Waikiki to Sea World*, Chicago: University of Chicago Press.

Dewsbury, J., Harrison, P., Rose, M. and Wylie, J. (2002) 'Enacting geographies: editorial introduction', *Geoforum*, 33(4): 437–40.

Digby, S. (2006) 'The casket of magic: home and identity from salvaged objects', *Home Cultures*, 3(2): 169–90.

Doorne, S. and Atejlevic, A. (2005) 'Tourism performance as metaphor: enacting backpacker travel in the Fiji Islands', in A. Jaworski and A. Pritchard (eds) *Discourse, Communication, and Tourism*, London: Multilingual Matters.

Douai, A. (2007) 'Tales of transgression or clashing paradigms: the Danish cartoon controversy and Arab media', *Global Media Journal*, 6(10): unpaginated.

du Gay, P., Hall, S., James, L., Mackey, H. and Negus, K. (1997) *Doing Cultural Studies: The Story of the Sony Walkman*, London: Sage.

Duim, V. D. R. (2007a) 'Tourismscapes: an actor network perspective', *Annals of Tourism Research*, 34(4): 961–76.

—— (2007b) 'Tourism, materiality and space', in I. Ateljevic, N. Morgan and A. Pritchard (eds) *The Critical Turn in Tourism Studies: Innovative Research Methodologies*, Oxford: Elsevier.

Duncan, J. and Gregory, D. (1999) 'Introduction', in J. Duncan and D. Gregory (eds) *Writes of Passage: Reading Travel Writing*, London: Routledge.

Duncan, J. and Lambert, D. (2003) 'Landscapes of home', in J. Duncan, N. Johnson and R. Schein (eds) *A Companion to Cultural Geography*, London: Blackwell.

Duval, T. Y. (2004a) 'Linking return visits and return migration among commonwealth Eastern Caribbean migrants in Toronto', *Global Networks*, 4(1): 51–8.

—— (2004b) 'Conceptualising return visits: a transnational perspective', in T. Coles and D. Timothy (eds) *Tourism, Diasporas and Space*, London: Routledge.

Edensor, T. (1998) *Tourists at the Taj: Performance and Meaning at a Symbolic Site*, London: Routledge.

—— (2000) 'Staging tourism: tourists as performers', *Annals of Tourism Research*, 27(2): 322–44.

—— (2001a) 'Performing tourism, staging tourism: (re)producing tourist space and practice', *Tourist Studies*, 1(1): 59–81.

—— (2001b) 'Walking in the British countryside: reflexivity, embodied practices and ways to escape', in P. Macnaghten and J. Urry (eds) *Bodies of Nature*, London: Sage.

—— (2006) 'Sensing tourist places', in C. Minca and T. Oaks (eds) *Travels in Paradox: Remapping Tourism,* Lanham, MD: Rowman & Littlefield.

Edwards, E. (ed.) (1992) *Photography and Anthropology*, New Haven, CT: Yale University Press.

—— (1996) 'Postcards: greetings from another world', in T. Selwyn (ed.) *The Tourist Image*, Chichester: John Wiley.

—— (1999) 'Photographs as objects of memory', in M. Kwint, C. Breward and J. Aynsley (eds) *Material Memories: Design and Evocation*, Oxford: Berg.

Edwards, E. and Hart, J. (2004) 'Introduction: photographs as objects', in E. Edwards (ed.) *Photographs Objects Histories: On the Materiality of Images*, London: Routledge.

Ek, R., Larsen, J., Hornskov, B. S. and Mansfeldt, O. (2008) 'A dynamic framework of tourist experiences: space-time and performances in the experience economy', *Scandinavian Journal of Hospitality and Tourism*, 8(2): 122–40.

Ellegaard, K. and Vilhelmson, B. (2004) 'Home as a pocket of local order: everyday activities and the friction of distance', *Geografiska Annaler Series*, B 86B(4): 281–96.

Erickson, K. (2004) 'Bodies at work: performing service in American restaurants', *Space and Culture*, 7(1): 76–89.

Eriksen, H. T. (2001) *Øyeblikkets Tyrani*, Oslo: Aschehoug.

Farrell, B. H. (1979) 'Tourism's human conflicts', *Annals of Tourism Research*, 6(2): 122–36.

Featherstone, M. (1992) 'The heroic life and everyday life', *Theory, Culture and Society*, 9: 159–82.

Felski, R. (1999) 'The invention of everyday life', *New Formations*, 39: 15–31.

Fennel, G. (1997) 'Local lives – distant ties: researching communities under globalized conditions', in J. Eade (ed.) *Living the Global City: Globalization as Local Process*, London: Routledge.

Fortier, A. (2000) *Migrant Belongings: Memory, Space, Identity*, Oxford: Berg.

Franklin, A. (2002) *Nature and Social Theory*, London: Sage.

—— (2003) *Tourism: An Introduction*, London: Sage.

—— (2007) 'Tourism as an ordering: towards a new ontology of tourism', *Tourist Studies*, 4(3): 277–301.

Franklin, A. and Crang, M. (2001) 'The trouble with tourism and travel theory', *Tourist Studies*, 1(1): 5–22.

Friedberg, S. (2001) 'On the trail of the global green bean: methodological considerations in multi-site ethnography', *Global Networks*, 1(4): 353–68.

Game, A. (1991) *Undoing the Social: Towards a Deconstructive Sociology*, Milton Keynes: Open University Press.

Gane, N. (2005) 'Radical post-humanism: Friedrich Kittler and the primacy of technology', *Theory, Culture & Society*, 22(3): 25–41.

Gergen, K. (2002) 'The challenge of absent presence', in J. Katz and M. Aakhus (eds) *Perpetual Contact: Mobile Communication, Private Talk, Public Performance*, Cambridge: Cambridge University Press.

Gibson, J. J. (1977) 'The theory of affordances', in R. Shaw and J. Brandsford (eds) *Perceiving, Acting and Knowing: Toward an Ecological Psychology*, Hillsdale, NJ: Lawrence Erlbaum Associates.

—— (1979) *The Ecological Approach to Visual Perception*, Boston: Houghton Mifflin.

—— (1982) (orig. 1938) 'A theoretical field-analysis of automobile-driving', in E. Reed and R. Jones (eds) *Reasons for Realism: Selected Essays by James J. Gibson*, Hillsdale and London: Lawrence Erlbaum Associates.

Gibson, S. (2007) 'Food mobilities', *Space and Culture*, 10(1): 4–21.

Giddens, A. (1984) *The Constitution of Society: Outline of the Theory of Structuration*, Berkeley: University of California Press.

—— (1990) *The Consequences of Modernity*, Cambridge: Polity Press.

Goffman, E. (1959) *The Presentation of Self in Everyday Life*, New York: Anchor Books.

—— (1963) *Behaviour in Public Places*, New York: Free Press.

—— (1989) 'On fieldwork' (transcribed and edited by Lyn H. Lofland), *Journal of Contemporary Ethnography*, 18(2): 123–32.

Goggin, G. (2006) *Cell Phone Culture: Mobile Technology in Everyday life*, London: Routledge.

Gosden, C. and Knowles, C. (2001) *Collecting Colonialism. Material Culture and Colonial Change*, Oxford: Berg.

Goss, J. (1993) 'Placing the market and marketing the place: tourist advertising of the Hawaiian Islands', *Environment and Planning D*, 11(6): 663–88.

—— (2004) 'The souvenir: conceptualizing the objects of tourist consumption', in A. Lew, A. Willams and C. M. Hall (eds) *A Companion to Tourism*, London: Blackwell.

Graves-Brown, P. (2000) 'Introduction', in P. Graves-Brown (ed.) *Matter, Materiality and Modern Culture*, London: Routledge.

Gregory, D. (1999) 'Scripting Egypt: Orientalism and the cultures of travel', in J. Duncan and D. Gregory (eds) *Writes of Passage: Reading Travel Writing*, London: Routledge.

—— (2004) *The Colonial Present: Afghanistan – Palestine – Iraq*, London: Blackwell.

—— (2005) 'Performing Cairo: Orientalism and the city of the Arabian Nights', in N. Alsayyad, I. Bierman and N. Rabat (eds) *Making Cairo Medieval*, Lanham, MD: Lexington Books.

—— (2007) 'Vanishing points', in D. Gregory and A. Pred (eds) *Violent Geographies: Fear, Terror and Political Violence*, London: Routledge.

Gregson, N. and Crewe, L. (2003) *Second-hand Cultures*, Oxford: Berg.

Gregson, N. and Rose, G. (2000) 'Taking Butler elsewhere: performativities, spatialities and subjectivities', *Environment and Planning D*, 18(4): 433–52.

Gregson, N., Metcalfe, A. and Crewe, L. (2007) 'Moving things along: the conduits and practices of divestment in consumption', *Transactions of the Institute of British Geographers*, 32(2): 187–200.

Gren, M. (2001) 'Time–geography matters', in J. May and N. Thrift (eds) *Timespace: Geographies of Temporality*, London: Routledge.

Grosrichard, A. (1998) *The Sultans Court: European Fantasies of the East*, London: Verso.

Gustafson, P. (2002) 'Tourism and seasonal retirement migration', *Annals of Tourism Research*, 29(3): 899–918.

Gye, L. (2007) 'Picture this: the impact of mobile camera phones on personal photographic practices', *Continuum*, 21(2): 279–88.

Hägerstrand, T. (1985) 'Time–geography: focus on the corporeality of man, society and individuals', in S. Aida (ed.) *The Science and Praxis of Complexity*, Tokyo: United Nations University.

Haldrup, M. (2004) 'Laid back mobilities', *Tourism Geographies*, 6(4): 434–54.

—— (2009) 'Banal tourism? Between cosmopolitanism and Orientalism', in P. Obrador, M. Crang and P. Travlou (eds) *Doing Tourism: Cultures of Mediterranean Mass Tourism*, Aldershot: Ashgate.

Haldrup, M. and Larsen, J. (2003) 'The family gaze', *Tourist Studies*, 3(1): 23–46.

—— (2006) 'Material cultures of tourism', *Leisure Studies*, 25(3): 275–89.

Haldrup, M., Koefoed, L. and Simonsen, K. (2006) 'Practical orientalism: bodies, everyday life and the construction of otherness', *Geografiske Annaler B*, 88(2): 173–84.

—— (2008) 'Practicing fear: encountering O/other bodies', in R. Pain and S. J. Smith (eds) *Fear: Critical Geopolitics and Everyday Life*, Aldershot: Ashgate.

Hall, C. M. (2005) 'Reconsidering the geography of tourism and contemporary mobility', *Geographical Research*, 43(3): 125–39.

Hall, C. M. and Müller, D. K. (eds) (2004) *Tourism, Mobility and Second Homes: Between Elite Landscape and Common Ground*, Clevedon: Channelview Press.

Halle, S. (1991) *Inside Culture: Art and Class in the American Home*, Chicago: Chicago University Press.

Hallman, C. B. and Penbow, P. M. S. (2007) 'Family leisure, family photography and zoos: exploring the emotional geographies of families', *Social and Cultural Geography*, 8(6): 871–88.

Hannam, K. (2006) 'Tourism and development III: performances, performativities and mobilities', *Progress in Development Studies*, 6(3): 243–49.

Hannam, K., Sheller, M. and Urry, J. (2006) 'Mobilities, immobilities and moorings', *Mobilities*, 1(1): 1–22.

Hannerz, U. (1996) *Transnational Connections: Culture, People, Places*, London: Routledge.

—— (2003) 'Being there . . . and there . . . and . . . there! Reflections on multi-site ethnography', *Ethnography*, 4(2): 201–16.

Hansen, R. (2006) 'The Danish cartoon controversy: a defence of liberal freedom', *International Migration*, 44(5): 7–16.

Harper, D. (2000) 'Remaining visual methods', in N. K Denzin and Y. S. Lincoln (eds) *Handbook of Qualitative Research*, London: Sage.

—— (2003) 'Framing photographic ethnography', *Ethnography*, 4(2): 241–66.

Harvey, D. (1989) *The Postmodern Condition*, London: Blackwell.

Hawass, Z. (1998) 'Site management: the response to tourism', *Museum International*, 50(4): 31–7.

Hashimoto, A. and Telfer, D. J. (2007) 'Geographical representations embedded within souvenirs in Niagara: the case of geographically displaced authenticity', *Tourism Geographies*, 9(2): 191–217.

Heidegger, M. (1993) 'Building dwelling thinking', in *Basic Writings*, London: Routledge.

Heller, A. (1984) *Everyday Life*, London: Routledge.

Hendry, J. (2001) *The Orient Strikes Back: A Global View of Cultural Display*, Oxford: Berg.

—— (2003) 'An ethnographer in the global arena: globography perhaps?', *Global Networks*, 3(4): 497–512.

Herbert, S. (2000) 'For ethnography', *Progress in Human Geography*, 24(4): 550–72.

Hetherington, K. (1997) 'In place of geometry: the materiality of place', in K. Hetherington and R. Munro (eds) *Ideas of Difference: Social Space and Labour of Division*, Oxford: Blackwell.

—— (2004) 'Secondhandedness: consumption, disposal, and absent presence', *Environment and Planning D*, 22(1): 157–73.

Hinchliffe, S. (1996) 'Technology, power, and space: the means and ends of geographies of technology', *Environment and Planning*, 14(6): 659–82.

—— (2003) ' "Inhabiting" – landscapes and natures', in K. Anderson, M. Domosh and S. Pile (eds) *Handbook of Cultural Geography*, London: Sage.

Hingham, B. (2002) 'Introduction: questioning everyday life', in B. Hingham (ed.) *The Everyday Life Reader*, London: Routledge.

Hirsch, M. (1997) *Family Frames: Photography, Narrative and Postmemory*, Cambridge: Harvard University Press.

Hitchcock, M. (2000) 'Introduction', in M. Hitchcock and K. Teague (eds) *Souvenirs: The Material Culture of Tourism*, Aldershot: Ashgate.

Hitchcock, M. and Teague, K. (eds) (2000) *Souvenirs: The Material Culture of Tourism*, Aldershot: Ashgate.

Hjorth, L. (2007) 'Snapshots of almost contact: the rise of camera phone practices and a case study in Seoul, Korea', *Continuum*, 21(2): 227–38.

Holland, P. (2001) 'Personal photography and popular photography', in L. Wells (ed.) *Photography: A Critical Introduction*, London: Routledge.

Holloway, L. and Hubbard, P. J. (2001) *People and Place: The Extraordinary Geographies of Everyday Life*, London: Prentice Hall.

Hughes, G. (1998) 'Tourism and the semiological realization of space', in G. Ringer (ed.) *Destinations: Cultural Landscapes of Tourism*, London: Routledge.

Hughes-Freeland, F. (ed.) (1998) *Ritual, Performance, Media* (ASA Monographs 35), London: Routledge.

Hutnyk, J. (1996) *The Rumour of Calcutta: Tourism, Charity and the Power of Representation*, London: Zed Books.

Ingold, T. (2000a) *The Perception of the Environment: Essays on Livelihood, Dwelling and Skill*, London: Routledge.

—— (2000b) 'Making culture and weaving the world', in P. Graves-Brown (ed.) *Matter, Materiality and Modern Culture*, London: Routledge.

Ingold, T. and Kurttila, T. (2001) 'Perceiving the environment in Finnish Lapland', in P. Macnaghten and J. Urry (eds) *Bodies of Nature*, London: Sage.

Iribas, J. M. (2000), 'Touristic urbanism', in W. Mass and ESARQ: *MVRDV, Costa Iberica, Upbeat to the Leisure City*, Barcelona: ACTAR.

Jackson, P. (1999) 'Commodity cultures: the traffic in things', *Transactions of the Institute of British Geographers*, 24(1): 95–108.

—— (2000) 'Rematerializing social and cultural geography', *Social and Cultural Geography*, 1(1): 9–14.

Jacobsen, S. K. J. (2003) 'The tourist bubble and the europeanisation of holiday travel', *Tourism and Cultural Change*, 1(1): 71–87.

Jarlöv, L. (1999) 'Leisure lots and summer cottages as places for people's own creative work', in D. Crouch (ed.) *Leisure/Tourism Landscapes: Practices and Geographical Knowledge*, London: Routledge.

Johanneson, G.T. and Bærenholdt, J.O. (forthcoming) 'Actor network theory/ networked geographies', in N. Thrift and R. Kitchen (eds) *International Encyclopedia of Human Geography*, Cheltenham: Elsevier.

Kabbani, R. (1994) *Imperial Fictions: Europe's Myth of Orient*, London: Pandora.

Kang, S. and Page, S. (2000) 'Tourism, migration and emigration: travel patterns of Korean-New Zealanders in the 1990s', *Tourism Geographies*, 2(3): 50–65.

Kasfir, S. L. (2004) 'Tourist aesthetics in the global flow: Orientalism and "warrior" theatre on the Swahili coast', *Visual Anthropology*, 17(3–4): 319–43.

Katz, J. (2000) *How Emotions Work*, Chicago: University of Chicago Press.

Kaufmann, V., Manfred, M. M. and Joye, D. (2004) 'Motility: mobility as social capital', *International Journal of Urban and Regional Research*, 28(4): 745–56.

Kirk, D., Sellen, A. Rother, C. and Wood, K. (2006) 'Understanding photowork', *CHI*, April: 761–70.

Koskinen, I. (2005) 'Pervasive image capture and sharing: methodological remarks', Pervasive Image Capture and Sharing Workshop, Ubiquitous Computing Conference, Tokyo.

Koskinen, I., Esko, K. and Lechtonen, K. T. (2002) *Professional Mobile Image*, Helsinki: IT Press.

Kuhn, A. (1995) *Family Secrets: Acts of Memory and Imagination*, London: Verso.

Kwint, M. (1999) 'Introduction: "the physical past"', in M. Kwint, C. Breward and J. Aynsley (eds) *Material Memories: Design and Evocation*, Oxford: Berg.

Kyle, G. and Chick, G. (2004) 'Enduring leisure involvement: the importance of personal relationships', *Leisure Studies*, 23(3): 243–66.

Lægaard, S. (2007) 'The cartoon controversy as a case of multicultural recognition', *Contemporary Politics*, 13(2): 147–64.

Larsen, J. (2001) 'Tourism mobilities and the travel glance: experiences of being on the move', *Scandinavian Journal of Hospitality and Tourism*, 1(2): 80–98.

—— (2003) 'Performing tourist photography', unpublished PhD thesis, Department of Geography, Roskilde University.

—— (2005) 'Families seen photographing: the performativity of tourist photography', *Space and Culture*, 8(3): 416–34.

—— (2006) 'Geographies of tourism photography: choreographies and performances', in J. Falkheimer and A. Jansson (eds) *Geographies of Communication: The Spatial Turn in Media Studies*, Gothenburg: Nordicom.

—— (2008) 'Practices and flows of digital photography: an ethnographic framework', *Mobilities*, 3(1): 141–60.

—— (2009) 'Goffman and the tourist gaze: a performativity approach to tourism mobilities', in M. H. Jacobsen (ed.) *Contemporary Goffman*, London: Routledge.

Larsen, J. and Jacobsen, H. M. (2009) 'Metaphors of mobility: social inequality on the move', in H. Maksim, T. Ohnmacht and M. Bergman (eds) *Mobility and Social Inequality*, Aldershot: Ashgate.

Larsen, J., Urry, J. and Axhausen, K. (2006) *Mobilities, Networks and Geographies*, Aldershot: Ashgate.

—— (2007) 'Networks and tourism: mobile social life', *Annals of Tourism Research*, 34(1): 244–62.

—— (2008) 'Coordinating face-to-face meetings in mobile network societies', *Information, Communication & Society*, 11(5): 640–58.

Lash, S. and Lury, C. (2007) *Global Culture Industry*, London: Polity.

Lash, S. and Urry, J. (1994) *Economies of Signs and Space*, London: Sage.

Latham, A. (2003) 'Research, performance, and doing human geography: some reflections on the diary-photograph, diary-interview method', *Environment and Planning A*, 35(11): 1993–2018.

Latour, B. (1991) 'Technology is society made durable', in J. Law (ed.) *A Sociology of Monsters: Essays on Power, Technology and Domination*, London: Routledge.

—— (1993) *We Have Never Been Modern*, Hemel Hempstead: Harvester Wheatsheaf.

—— (2000) 'The Berlin key or how to do words with things', in P. Graves-Brown (ed.) *Matter, Materiality and Modern Culture*, London: Routledge.

—— (2005) *Reassembling the Social: An introduction to Actor Network-Theory*, Oxford: Oxford University Press.

Laustsen, C. (2008) 'The camera as a weapon: on Abu Ghraib and related matters', *Journal for Cultural Research*, 12(2): 123–42.

Law, J. (1994) *Organizing Modernity*, Oxford: Basil Blackwell.

Lefebvre, H. (1991) *Critique of Everyday Life*, London: Verso.

Lew, A., Hall, C. M. and Williams, A. (eds) (2004) *A Companion to Tourism*, London: Blackwell.

Licoppe, C. (2004) '"Connected" presence: the emergence of a new repertoire for managing social relationships in a changing communication technoscape', *Environment and Planning D*, 22: 135–56.

Liniado, M. (1996) *Car Culture and Countryside Change*, Cirencester: National Trust.

Lister, M. (1995) 'Introductory essay', in M. Lister (ed.) *The Photographic Image in Digital Culture*, London: Routledge.

—— (2001) 'Photography in the age of electronic imaging', in L. Wells (ed.) *Photography: A Critical Introduction*, London: Routledge.

—— (2007) 'A sack in the sand: photography in the age of information', *Convergence*, 13(3): 251–74.

Littlewood, I. (2001) *Sultry Climates: Travel and Sex since the Grand Tour*, London: John Murray.

Lodge, D. (1991) *Paradise News*, London: Penguin.

Löfgren, O. (1999) *On Holiday: A History of Vacationing*, Berkeley: University of California Press.

Lorimer, H. (2005) 'Cultural geography: the busyness of being "more-than-representational"', *Progress in Human Geography*, 29(3): 83–94.

Lury, C. (1997) 'The objects of travel', in C. Rojek and J. Urry (eds) *Touring Cultures. Transformations of Travel and Theory*, London: Sage.

—— (1998) *Prosthetic Culture: Photography, Memory and Identity*, London: Routledge.

McCabe, S. (2002) 'The tourist experience and everyday life', in G. Dann (ed.) *The Tourist as a Metaphor of the Social World*, Wallingford: CABI.

MacCannell, D. (1976/1999) *The Tourist: A New Theory of the Leisure Class*, New York: Schocken Books.

McGlone, F., Park, A. and Roberts, C. (1999) 'Kinship and friendship: attitudes and behaviour in Britain 1986–1995', in S. McRae (ed.) *Changing Britain: Families and Households in the 1990s*, Oxford: Oxford University Press.

Macnaghten, P. and Urry, J. (2001) 'Bodies of nature: introduction', in P. Macnaghten and J. Urry (eds) *Bodies of Nature*, London: Sage.

Malam, L. (2004) 'Performing masculinity on the Thai beach scene', *Tourism Geographies*, 6(4): 455–71.

Marcus, G. E. (1995) 'Ethnography in/of the world system: the emergence of multi-sited ethnography', *Annual Review of Anthropology*, 24: 95–117.

—— (1998) *Ethnography through Thick and Thin*, Princeton: Princeton University Press.

Markus, G. (2001) 'Walter Benjamin or: the commodity as phantasmagoria', *New German Critique*, 83: 3–42.

Markwick, M. (2001) 'Postcards from Malta: image, consumption, context', *Annals of Tourism Research*, 28(2): 417–38.

Mason, J. (2004) 'Managing kinship over long distances: the significance of "the visit"', *Social Policy & Society*, 3(3): 421–9.

Massey, D. (1994) *Space, Place and Gender*, Cambridge: Polity Press.

—— (1995) 'The conceptualisation of place', in D. Massey and P. Jess (eds) *A Place in the World*, Oxford: Open University Press.

Merrington, P. (2001) 'A staggered Orientalism: the Cape-to-Cairo imaginary', *Poetics Today*, 22(2): 323–59.

Merleau-Ponty, M. (1962) *Phenomenology of Perception*, London: Routledge and Kegan Paul.

Michael, M. (2000) *Reconnecting Culture: Technology and Nature – From Society to Heterogeneity*, London: Routledge.

Miller, D. (1997) *Capitalism: An Ethnographic Approach*, Oxford: Berg.

—— (1998) 'Why some things matter', in D. Miller (ed.) *Material Culture: Why Some Things Matter*, Chicago: University of Chicago Press.

—— (ed.) (2001a) *Home Possessions: Material Culture behind Closed Doors*, Oxford: Berg.

—— (ed.) (2001b) *Car Cultures*, London: Berg.

—— (2001c) 'Behind closed doors', in D. Miller (ed.) *Home Possessions: Material Culture behind Closed Doors*, Oxford: Berg.

—— (2006) 'Things that bright up the place', *Home Cultures*, 3(3): 235–49.

Miller, D. A. and Edwards, K. W. (2007) 'Give and take: A study of consumer photo-sharing culture and practice', *CHI*, 28 April–3 May.

Miller, D. and Slater, D. (2000) *The Internet*, London: Berg.

Miller, D., Jackson, P., Thrift, N., Holbrook, B. and Rowlands, M. (1998) *Shopping, Place and Identity*, London: Routledge.

Minca, C. and Oakes, T. (2006) *Travels in Paradox*, Boulder, CO: Rowman and Littlefield.

Mitchell, T. (1988) *Colonizing Egypt*, Cambridge: Cambridge University Press.

Molz, J. G. (2004) 'Playing online and between the lines: round-the-world websites as virtual places to play', in M. Sheller and J. Urry (eds) *Tourism Mobilities: Places to Play, Places in Play*, London: Routledge.

—— (2006a) ' "Watch us wander": mobile surveillance and the surveillance of mobility', *Environment and Planning A*, 38(2): 377–93.

—— (2006b) 'Cosmopolitan bodies: fit to travel and travelling to fit', *Body and Society*, 12(3): 1–21.

—— (2008) 'Global abode: home and mobility in narratives of round-the-world travel', *Space and Culture*, 11(4): 325–42.

Mordue, T. (2001) 'Performing and directing resident/tourist cultures in Heartbeat country', *Tourist Studies*, 1(3): 233–52.

Morgan, N. and Pritchard, N. (2005) 'On souvenirs and metonymy: narratives of memory, metaphor and materiality', *Tourist Studies*, 5(1): 29–53.

Morrisey (1994) *Vauxhall and I* (album), Warner Bros/WEA.

Murdoch, J. (1996) 'Inhuman/nonhuman/human: actor network theory and the prospects for a nondualistic and symmetrical perspective on nature and society', *Environment and Planning D: Society and Space*, 15: 731–56.

Nash, C. (2000) 'Performativity in practice: some recent work in cultural geography', *Progress in Human Geography*, 24(4): 653–64.

Nava, M. (2002) 'Cosmopolitan modernity: everyday imaginaries and the register of difference', *Theory, Culture and Society*, 1(2): 81–99.

Nazia, A. and Holden, A. (2006) 'Post-colonial Pakistani mobilities: the embodiment of the "myth of return" in tourism', *Mobilities*, 1(3): 217–42.

Norman, A. D. (1999) 'Affordances, conventions, and design', *Interactions*, May–June: 38–42.

Obrador-Pons, P. (2003) ' "Being-on-holiday": tourist dwelling, bodies and place', *Tourist Studies*, 3(1): 47–67.

—— (2007) 'A haptic geography of the beach: naked bodies, vision and touch', *Social and Cultural Geography*, 8(1): 123–41.

Obrador-Pons, P., Travlou, P. and Crang, M. (eds) (2010, forthcoming) *Cultures of Mass Tourism: Doing the Mediterranean in the Age of Banal Mobilities*, Aldershot: Ashgate.

O'Connor, K. (2006) 'Kitsch, tourist art, and the little grass shack in Hawaii', *Home Cultures*, 3(3): 251–71.

O'Reilly, K. (2003) 'When is a tourist? The articulation of tourism and migration in Spain's Costa del Sol', *Tourist Studies*, 3(3): 301–17.

Osborne, P. (2000) *Travelling Light: Photography, Travel and Visual Culture*, Manchester: Manchester University Press.

Ousby, I. (1990) *The Englishman's England: Taste, Travel and the Rise of Tourism*, Cambridge: Cambridge University Press.

Parinello, G. L. (2001) 'The technological body in tourism, research and praxis', *International Sociology*, 16(2): 205–19.

Pels, D., Hetherington, K. and Vandenberg, F. (2002) 'The status of the object: performances, mediations, and techniques', *Theory, Culture and Society*, 19(5/6): 1–21.

Perkins, H. and Thorns, D. (2001) 'Gazing or performing? Reflections on Urry's tourist gaze in the context of contemporary experiences in the Antipodes', *International Sociology*, 16(2): 185–204.

Philo, C. (2000) 'More words, more worlds: reflections on the "cultural turn" and human geography', in I. Cook, D. Crouch, S. Naylor and J. Taylor (eds) *Cultural Turns/ Geographical Turns*, London: Prentice Hall.

Pinch, T. J. and Bijker, E. W. (1984) 'The social construction of facts and artefacts: or how the sociology of science and the sociology of technology might benefit each other', *Social Studies of Science*, 14(3): 399–441.

Pink, S. (2001) *Doing Visual Ethnography: Images, Media and Representation in Research*, London: Sage.

Poster, M. (1999) 'Underdetermined', *New Media and Society*, 1(1): 12–18.

Preston-Whyte, R. (2002) 'Constructions of surfing space at Durban, South Africa', *Tourism Geographies*, 4(3): 307–28.

Putnam, D. R. (2001) *Bowling Alone: The Collapse and Revival of American Community*, New York: Simon & Schuster.

Quinn, B. (2007) 'Performing tourism: Venetian residents in focus', *Annals of Tourism Research*, 34(2): 458–76.

Radley, A. (1990) 'Artefacts, memory, and a sense of place', in D. Middleton and D. Edwards (eds) *Collective Remembering*, London: Sage.

Retzinger, J. (1998) 'Framing the tourist gaze: railway journeys across Nebraska, 1866–1906', *GPQ*, 18: 213–26.

Ringer, G. (1998) *Destinations: Cultural Landscapes of Tourism*, London: Routledge.

Ritchie, B. W., Burns, P. and Palmers, C. (eds) (2005) *Tourism Research Methods: Integrating Theory with Practice*, Wallingford: CABI.

Ritzer, G. (1997) *The McDonaldization Thesis: Explorations and Extensions*, London: Routledge.

Ritzer, G. and Liska, A. (1997) ' "McDisneyization" and "post-tourism": complementary perspectives on contemporary tourism', in C. Rojek and J. Urry (eds) *Touring Culture: Transformations of Travel and Theory*, London: Routledge.

Robins, B. (1998) 'Actually existing cosmopolitanism', in P. Cheah and B. Robbins (eds) *Cosmopolis: Thinking and Feeling beyond the Nation*, Minneapolis: University of Minnesota Press.

Robins, K. (1991) 'Into the image: visual technologies and vision cultures', in P. Wombell (ed.) *Photovideo: Photography in the Age of the Computer*, New York: Paul & Co.

Rodaway, P. (1994) *Sensuous Geographies: Body, Sense and Place*, London: Routledge.

Rojek, C. (1993) *Ways of Escape*, London: Sage.

—— (1995) *Decentring Leisure: Rethinking Leisure Theory*, London: Sage.

Rojek, C. and Urry, J. (1997) 'Introduction', in C. Rojek and J. Urry (eds) *Touring Cultures: Transformations of Travel and Theory*, London: Routledge.

Rose, G. (2001) *Visual Methodologies: An Introduction to the Interpretation of Visual Materials*, London: Sage.

—— (2003) 'Family photographs and domestic spacings: a case study', *Transactions of the Institute of British Geographers*, 28(1): 5–18.

Rose, M. (2002) 'Landscapes and labyrinths', *Geoforum*, 33(4): 455–67.

Ryan, J. (1997) *Picturing Empire: Photography and the Visualisation of the British Empire*, London: Reaktion Books.

Sachs, W. (1992) *For Love of the Automobile*, Berkeley: University of California Press.

Said, E. (1995) *Orientalism: Western Conceptions of the Orient*, London: Penguin.

—— (2004) 'Orientalism once more', *Development and Change*, 35(5): 869–79.

Schechner, R. (2006) *Performances Studies: An Introduction*, London: Routledge.

Scheibe, K. E. (1986) 'Self-narratives and adventure', in T. R. Sarbin (ed.) *Narrative Psychology*, New York: Praeger.

Schieffelin, E. (1998) 'Problematizing performance', in F. Hughes-Freeland (ed.) *Ritual, Performance, Media* (ASA Monographs 35), London: Routledge.

Schivelbusch, W. (1979) *The Railway Journeys: Trains and Travel in the 19th Century*, Oxford: Blackwell.

Selwyn, T. (1996) *The Tourist Image*, Chichester: John Wiley & Sons.

Sheller, M. (2003) *Consuming the Caribbean*, London: Routledge.

—— (2004) 'Automotive emotions: feeling the car', *Theory, Culture and Society*, 21(4–5): 221–42.

Sheller, M. and Urry, J. (eds) (2004) *Tourism Mobilities: Places to Play, Places in Play*, London: Routledge.

Shenkav-Keller, S. (1993) 'The Israeli souvenir: its text and context', *Annals of Tourism Research*, 20: 182–96.

Shields, R. (1991) *Places on the Margins: Alternative Geographies of Modernity*, London: Routledge.

—— (2004) 'Surfing: global space or dwelling in the waves', in M. Sheller and J. Urry (eds) *Tourism Mobilities: Places to Play, Places in Play*, London: Routledge.

Shiner, L. (1994) ' "Primitive fakes", tourist art and the ideology of authenticity', *Journal of Aesthetics and Art Criticism*, 52(2): 225–34.

Shove, E., Watson, M., Hand, M. and Ingram, J. (2007) *Design of Everyday Life*, Oxford: Berg.

Simmel, G. (1950) 'The metropolis and mental life', in K. H. Wolff (ed.) *The Sociology of George Simmel*, New York: Free Press.

—— (1997a) 'The sociology of sociability', in D. Frisby and M. Featherstone (eds) *Simmel on Culture*, London: Sage.

—— (1997b) 'Sociology of the meal', in D. Frisby and M. Featherstone (eds) *Simmel on Culture*, London: Sage.

Simonsen, K. (1999) 'Difference in human geography: travelling through Anglo-Saxon and Scandinavian discourses', *European Planning Studies*, 7(1): 9–24.

—— (2003) 'On being "in-between": social and cultural geography in Denmark', *Social & Cultural Geography*, 4(2): 255–68.

Skrbis, Z., Kendal, G. and Woodward, I. (2004) 'Locating cosmopolitanism', *Theory, Culture & Society*, 21(6): 115–36.

Slater, D. (1995) 'Domestic photography and digital culture', in M. Lister (ed.) *The Photographic Image in Digital Culture*, London: Routledge.

Slyomovics, S. (1989) 'Cross-cultural dress and tourist performance in Egypt', *Performing Arts Journal*, 11(3): 139–48.

Smith, L. (1998) *The Politics of Focus: Women, Children, and Nineteenth-Century Photography*, London: St. Martin's Press.

Smith, V. L. (1977) 'Introduction', in V. L. Smith (ed.) *Hosts and Guests: The Anthropology of Tourism*, Philadelphia: University of Pennsylvania Press.

Sontag, S. (1978) *On Photography*, New York: Farrar, Straus & Giroux.

—— (2004a) 'Regarding the torture of others', *New York Times*, 23 May.

—— (2004b) *Regarding the Pain of Others*, London: Penguin.

Stewart, S. (1993) *On Longing: Narratives of the Miniature, the Gigantic, the Souvenir, the Collection*, London: Duke University Press.

Sutton, R. C. (2004) 'Celebrating ourselves: the family reunion rituals of African Caribbean transnational families', *Global Networks*, 4(3): 243–58.

Suvantola, J. (2002) *Tourist's Experience of Place*, Burlington: Ashgate.

Szerszynski, B. and Urry, J. (2006) 'Visuality, mobility and the cosmopolitan: inhabiting the world from afar', *British Journal of Sociology*, 57(1): 113–31.

Szerszynski, B., Heim, W. and Waterton, C. (2003) 'Introduction', in B. Szerszynski, W. Heim and C. Water (eds) *Nature Performed: Environment, Culture and Performance*, London: Blackwell.

Thompson, J. B. (1989) *Media and Modernity*, Cambridge: Polity.

Thrift, N. (1996) *Spatial Formations*, London: Sage.

—— (1997) 'The still point: resistance, expressive embodiment and dance', in S. Pile and M. Keith (eds) *Geographies of Resistance*, London: Routledge.

—— (1999) 'Steps to an ecology of place', in D. Massey and P. Sarre (eds) *Human Geography Today*, Cambridge: Polity.

—— (2004a) 'Driving the city', *Theory, Culture and Society*, 21(4–5): 41–59.

—— (2004b) 'Intensities of feeling: towards a spatial politics of affect', *Geografiska Annaler*, 86B(1): 57–78.

—— (2007) *Non-representational Theories*, London: Routledge.

Thrift, N. and Dewsbury, J.-D. (2000) 'Dead geographies – and how to make them live', *Environment and Planning D*, 18(4): 411–32.

Tolia-Kelly, D. (2004a) 'Locating processes of identification: studying the precipitates of re-memory through artefacts in the British Asian home', *Transactions of the Institute of British Geographers*, 29(3): 314–29.

—— (2004b) 'Materializing post-colonial geographies: examining the textural landscapes of migration in the British Asian home', *Geoforum*, 35: 675–88.

Travel Trends (2004) *Travel Trends 2003: A Report on the International Passenger Survey*, London: National Statistics.

Tuan, Y.-F. (1996) 'Space and place: humanistic perspective', in J. Agnew, D. N. Livingstone and A. Rogers (eds) *Human Geography: An Essential Anthology*, Oxford: Blackwell.

Turner, B. S. (1994) *Orientalism, Postmodernism and Globalism,* London: Routledge.

Tzanelli, R. (2007) *The Cinematic Tourist: Explorations in Globalization, Culture and Resistance*, London: Routledge.

Uriely, N. (2005) 'The tourist experience: conceptual developments', *Annals of Tourism Research*, 32(3): 199–216.

Urry, J. (1990) *The Tourist Gaze*, London: Sage.

—— (1995) *Consuming Places*, London: Routledge.

—— (2000) *Sociology beyond Society: Mobilities for the Twenty-first Century*, London: Routledge.

—— (2002a) *The Tourist Gaze*, 2nd edn, London: Sage.

—— (2002b) 'Mobility and proximity', *Sociology*, 36(2): 255–74.

—— (2003) 'Social networks, travel and talk', *British Journal of Sociology*, 54(2): 155–76.

—— (2004) 'Connections', *Environment and Planning D*, 22(1): 27–37.

—— (2007) *Mobilities*, Cambridge: Polity Press.

Van Dijck, J. (2008) 'Digital photography: communication, identity, memory', *Visual Communication*, 7(1): 57–76.

Van House, N. (2007) 'Flickr and public image-sharing: distant closeness and photo exhibition', *CHI*, 28 April–3 May: 2717–22.

Van House, N., Davis, M., Takhteyev, Y., Ames, M. and Finn, F. (2004) 'The social uses of personal photography: methods for projecting future imaging applications', retrieved 3 December 2005 (http://www.sims.berkeley.edu/~vanhouse/pubs.htm).

Van House, N., Davis, M., Ames, M., Finn, M. and Viswanathan, V. (2005) 'The uses of personal networked digital imaging: an empirical study of cameraphone photos and sharing', *CHI*, 2–7 April: 1853–6.

Van Leeuwen, T. and Jewitt, C. (2001) *Handbook of Visual Analysis*, London: Sage.

Van Maanen, J. (1988) *Tales from the Field: On Writing Ethnography*, Chicago: University of Chicago Press.

Veijola, S. (2006) 'Heimat tourism in the countryside: paradoxical sojourns to self and place', in C. Minca and T. Oakes (eds) *Travels in Paradox: Remapping Tourism*, Lanham, MD: Rowman & Littlefield.

Villi, M. (2007) 'Mobile visual communication: photo messages and camera phone photography', *Nordicom Review*, 28(1): 49–62.

Wagner, J. (2002) 'Contrasting images, complementary trajectories: sociology, visual sociology and visual research', *Visual Studies*, 17(2): 160–71.

Waitt, G. and Head, L. (2002) 'Postcards and frontier mythologies: sustaining views of the Kimberley as timeless', *Environment and Planning D: Society and Space*, 20(3): 319–44.

Wallendorf, M. and Arnould, E. J. (1988) ' "My favourite things": a cross-cultural inquiry into object attachment, possessiveness, and social linkage', *Journal of Consumer Research*, 14: 531–47.

Wang, N. (1999) 'Rethinking authenticity in tourism experience', *Annals of Tourism Research*, 26: 349–70.

Warde, A. (2005) 'Consumption and theories of practice', *Journal of Consumer Culture*, 5(2): 131–53.

Watson, M. (2008) 'The materials of consumption', *Journal of Consumer Culture*, 8(1): 5–10.

Weave, A. (2005) 'Interactive service work and performative metaphors: the case of the cruise industry', *Tourist Studies*, 5(1): 5–27.

Wellman, B. (2001) 'Physical place and cyberplace: the rise of personalised networking', *International Journal of Urban and Regional Research*, 25(2): 227–52.

—— (2002) *Little Boxes, Glocalization, and Networked Individualism* (available at http://www.chass.utoronto.ca/-wellman).

Wellman, B., Hogan, B., Berg, K., Boase, J., Carrasco, J. A., Côté, R., Kayahara, J., Kennedy, T. and Tran, P. (2006) 'Connected lives: the project', in P. Purcell (ed.) *The Networked Neighborhood*, Berlin: Springer.

Wells, L. (2003) 'General introduction', in L. Wells (ed.) *The Photography Reader*, London: Routledge.

West, B. (2006) 'Consuming national themed environments abroad: Australian working holidaymakers and symbolic national identity in "Aussie" theme pubs', *Tourist Studies*, 6(2): 139–55.

Whatmore, S. (1999) 'Hybrid geographies: rethinking the "human" in human geography', in D. Massey, J. Allen and P. Sarre (eds) *Human Geography Today*, London: Routledge.

—— (2002) *Hybrid Geographies: Natures, Cultures, Spaces*, London: Sage.

White, B. P. and White, R. W. (2005) 'Keeping connected: travelling with the telephone', *Convergence*, 11(2):102–12.

—— (2007) 'Home and away: tourists in a connected world', *Annals of Tourism Research*, 34(1): 88–104.

Williams, D. and Kaltenborn, B. (1999) 'Leisure places and modernity: the use and meaning of recreational cottages in the Norway and USA', in D. Crouch (ed.) *Leisure/Tourism Geographies: Practices and Geographical Knowledge*, London: Routledge.

Williams, M. A. and Hall, C. M. (2000) 'Tourism and migration: new relationships between production and consumption', *Tourism Geographies*, 2(3): 5–27.

Williams, M. A., King, R., Warnes, A. and Patters, G. (2000) 'Tourism and international retirement migration: new forms of an old relationship in Southern Europe', *Tourism Geographies*, 2(3): 28–49.

Wilson, A. (1992) *The Culture of Nature*, Oxford: Blackwell.

Wittel, A. (2000) 'Ethnography on the move: from field to net to internet', *Forum Qualitative Sozialforschung/Forum: Qualitative Social Research* 1(1) (http://qualitative-research.net/fqs).

World Tourist Organization (WTO) (2008) *Tourism Highlights* (http://www.unwto.org/facts/eng/pdf/highlights/UNWTO_Highlights08_en_HR.pdf).

Wren, K. (2001) 'Cultural racism: something rotten in the state of Denmark?', *Social and Cultural Geography*, 2(2):141–62.

Wylie, J. (2005) 'A single day's walking: narrating self and landscape on the South West Coast Path', *Transactions of the Institute of British Geographers*, 30: 234–47.

Žižek, S. (2008) *Violence*, London: Profile Books.

Index

157–8; home ethnographies, research methodology and use of 158, 161, 162, 163, 166–7, 168, 174, 176; life after tourism 154–6, 157–67; life cycle of 162–7; material objects as 161; meaningful souvenirs 158–61; souvenir work 154–5; tourist souvenirs 51; trivial souvenirs 156–8; use-value souvenirs 166–7

space of flows 43

spacial closeness, social remoteness 22–3

Spain 51

spatio-temporal reality 72–3

Stephonson, Tom 73

Stewart, S. 159, 161

strangeness 75

sunbathing 109–10

surfboarding 72–3

Sutton, R.C. 30

Suvantola, J. 20

Szerszynski, B. and Urry, J. 24, 187

Szerszynski, B. *et al.* 62

take-off on holiday 95–7

Tales of the Field (van Mannen, J.) 94–5

taskscapes 65–6

teamwork in photography 129–30, 144–5

technical determinism 122

technological change, transformation of tourism and 14–15

technological performance: access to communication technologies and 44; digital photography and convergence with new media technologies 130–1, 132; tourist performance and 11–12, 63

television and DVDs 110–12

theatrical choreography of Orientalism 85–6

theorization of photography 123–5

Thomas Cook & Sons 78

Thompson, J.B. 42

Thrift, N. 7, 9, 10, 12, 38, 58, 72, 74, 124, 125

Thrift, N. and Dewsbury, J.-D. 9, 11, 125

time–space distanciation 24–5, 42

Tivoli Gardens, Copenhagen 78

Tolia-Kelly, D. 163, 175

tourism: accessibility of 1, 30; corporeal travel (tourism) 16, 18, 24; cultural accounts of 3; cultures of, materialization of 59–61; embodied tourism 10, 13–14; enclosure in 108; experiences of 3–7; fragility of tourists' imaginations 18; local tourism performance 186; mundane

routines, tourism as escape from 20–1; overnight stays, defining tourism on basis of 31; relations between tourists 106–9; staging of authenticity in 6–7; tourism mobilities and cosmopolitan culture 194–5; tourism production 47; tourism writing 5; 'tourist gaze' 3, 20–4; touristification of everyday life 2; unpredictability of travel and 18; VFR (visiting friends and relatives) tourism 30–1; vulnerability of travel and 18; *see also* cosmopolitanism, tourist mobilities and; consumption of tourism; de-exoticization of tourist travel; digital photography; everyday lives on holiday; life after tourism; Orientalism, Orient and; tourist performance

tourism studies: circuit of performance model 4; hermenuetic circle of sight-seeing 4; passive tourism 3–4; performance turn in 2, 3–7; shaping tourist places 4

'tourismscapes' 42

tourist bubbles 32

The Tourist Gaze (Urry, J.) 20–4

tourist performance 58–74; affordance, Gibson's notion of 2, 7–9, 62; authenticity, staging in 6–7; automobilization 72; bazaars 69–70; camera obscura 63; circuit of performance model 4; collaborative nature of 26–8; collective performance 6, 26–8; consumption and production in 4, 5; corporeal mobility 59–61, 62, 65; creativity and improvisation in 13; culture as relational achievement between humans and non-humans 60–1; embodied tourism 62–3; fragile places 64–6; guidebook influences on 5; hybrid geographies 58–9; hybridized mobility 71–3, 74; imaginative mobilities 64; interactions with space and objects 61; landscape experience 73; material cultures 58–74; material environments 63–4; material landscapes, sensual involvement with 66; material semiotics 62; materiality and culture, artificial dualism between 59–61; materialization of 7–13, 18–19; memories of places 64; multidimensionality 62; mundane and collective 26–8; non-representational theory 9–10, 11; object mobilities 65; Orient, Western